Hammoud Aljoumaa

Development of a Self-Learning Approach

Hammoud Aljoumaa

Development of a Self-Learning Approach

Applied to Pattern Recognition and Fuzzy Control

Südwestdeutscher Verlag für Hochschulschriften

Impressum / Imprint

Bibliografische Information der Deutschen Nationalbibliothek: Die Deutsche Nationalbibliothek verzeichnet diese Publikation in der Deutschen Nationalbibliografie; detaillierte bibliografische Daten sind im Internet über http://dnb.d-nb.de abrufbar.

Alle in diesem Buch genannten Marken und Produktnamen unterliegen warenzeichen-, marken- oder patentrechtlichem Schutz bzw. sind Warenzeichen oder eingetragene Warenzeichen der jeweiligen Inhaber. Die Wiedergabe von Marken, Produktnamen, Gebrauchsnamen, Handelsnamen, Warenbezeichnungen u.s.w. in diesem Werk berechtigt auch ohne besondere Kennzeichnung nicht zu der Annahme, dass solche Namen im Sinne der Warenzeichen- und Markenschutzgesetzgebung als frei zu betrachten wären und daher von jedermann benutzt werden dürften.

Bibliographic information published by the Deutsche Nationalbibliothek: The Deutsche Nationalbibliothek lists this publication in the Deutsche Nationalbibliografie; detailed bibliographic data are available in the Internet at http://dnb.d-nb.de.

Any brand names and product names mentioned in this book are subject to trademark, brand or patent protection and are trademarks or registered trademarks of their respective holders. The use of brand names, product names, common names, trade names, product descriptions etc. even without a particular marking in this works is in no way to be construed to mean that such names may be regarded as unrestricted in respect of trademark and brand protection legislation and could thus be used by anyone.

Coverbild / Cover image: www.ingimage.com

Verlag / Publisher:
Südwestdeutscher Verlag für Hochschulschriften
ist ein Imprint der / is a trademark of
AV Akademikerverlag GmbH & Co. KG
Heinrich-Böcking-Str. 6-8, 66121 Saarbrücken, Deutschland / Germany
Email: info@svh-verlag.de

Herstellung: siehe letzte Seite /
Printed at: see last page
ISBN: 978-3-8381-3715-5

Zugl. / Approved by: Duisburg, Duisburg-Essen Uni, Uni, Diss., 2012

Copyright © 2013 AV Akademikerverlag GmbH & Co. KG
Alle Rechte vorbehalten. / All rights reserved. Saarbrücken 2013

Dedicated to my Lord and my Guide

My parents

My wife

Acknowledgements

First of all, I would like to thank my supervisor Univ.-Prof. Dr.-Ing. Dirk Söffker for his great support, encouragement, and help. Without him, this work would never initiate nor finish.

I would also like to thank Prof. Dr.-Ing. Uwe Maier for his effort being the co-reviewer for my thesis.

I would like to thank all my long time working partners at the Chair of Dynamics and Control; especially, Prof. Dr.-Ing. Yousef Al-Sweiti, Mahmud-Sami Saadawia, Lou'i Al-Shrouf, Dorra Baccar, Mustafa Turki Hussein, Dr.-Ing. Markus Özbek, Dr.-Ing. Dennis Gamrad and Marcel Langer for their help and support and our secretaries Yvonne Vengels and Doris Schleithoff for their help with the administration. I wish them all the best in their future life.

Special thanks are given to my wife who gives me always support, love, and strength. Finally but mostly I would like to thank my brothers and sisters and especially my parents whom without their love and support I would never be able to conduct this work.

Essen, 27. Oktober 2012   Hammoud Aljoumaa

Contents

Nomenclature   **VIII**

1 Introduction   1
    1.1 State of art 3
        1.1.1 Non-fuzzy-based pattern recognition approaches 3
        1.1.2 Pattern recognition approaches of time series data 4
        1.1.3 Fuzzy-based pattern recognition approaches 6
    1.2 Motivation 6
        1.2.1 Representation form for fuzzy rules 7
        1.2.2 Fuzzy partition technique 8
        1.2.3 Fuzzy rule extraction 9
        1.2.4 Criterion for determining the number of fuzzy rules 9
        1.2.5 Fuzzy rule selection 10
    1.3 Organization of the work 11

2 Pattern recognition and fuzzy rule-based systems   12
    2.1 Pattern recognition systems 12
        2.1.1 Definition 12
        2.1.2 Basic concepts 12
        2.1.3 Design procedure 13
        2.1.4 Data collection 13
        2.1.5 Feature extraction 14
        2.1.6 Feature selection 15
        2.1.7 Classifier design 16
        2.1.8 System evaluation 21
    2.2 Fuzzy rule-based systems 21
        2.2.1 Basic concepts 21
        2.2.2 Fuzzy rule-based system 25
        2.2.3 Application areas of fuzzy rule-based systems 29

3 Adaptive Fuzzy-Based Approach (AFBA)   39
    3.1 Used terminology 39
    3.2 General structure 39
    3.3 Feature extraction based on a sliding window 40
    3.4 Embedded fuzzy-based modeling 41
        3.4.1 Homogeneity-oriented vector 42
        3.4.2 Fuzzy partition process 43
        3.4.3 Adaptation of the boundary parameters in the feature space 44
        3.4.4 Fuzzy rule generation and extraction 48
    3.5 Fuzzy model 48
    3.6 Classification process 49

4	**Experimental validation of the new approach**		**51**
	4.1	Implementation of the (AFBA) approach based on the benchmark datasets	51
		4.1.1 Time series/streamed datasets	51
		4.1.2 Structured datasets	52
	4.2	Application of AFBA approach to tribological system	56
		4.2.1 Test rig	57
		4.2.2 Problem statement and hypothesis	57
		4.2.3 Training and modeling phase	58
		4.2.4 Representations spaces	59
		4.2.5 Testing and evaluation phase	62
5	**Summary and conclusions**		**66**
	5.1	Scientific contribution	67
	5.2	Future aspects	68
References			**69**

List of Figures

1.1	Examples of the modern complex systems	1
1.2	Example of procedure of pattern recognition process	2
2.1	Basic steps in the design procedure	14
2.2	Neuron model	18
2.3	Activation functions	18
2.4	Multilayer neural network	19
2.5	Optimal hyperplane and maximal margin in feature space	20
2.6	Forms of the standard membership functions (triangular, trapezoidal, and generalized Bell-shaped)	22
2.7	Forms of the standard membership functions (Gaussian, Z-shaped, and Sigmoidal)	23
2.8	Specifications of the membership function	24
2.9	Basic structure of the fuzzy-based system	27
2.10	Mamdani fuzzy control as the binary fuzzy rule-based system	30
2.11	Takagi-Sugeno fuzzy control as the binary fuzzy rule-based system	31
2.12	Classical PID controller and incremental variant of the fuzzy PID controller	31
2.13	Possible structure of a neuro-fuzzy/fuzzy-neuro controller	32
2.14	General structure of adaptive fuzzy control systems	33
2.15	Types of the fuzzy neuron based on the operations	35
3.1	Basic structure of AFBA	40
3.2	Feature extraction process using sliding windows	41
3.3	Example to explain the homogeneity-oriented vector	42
3.4	Triangular fuzzy membership function	43
3.5	Statistical characteristics-based fuzzy partition technique within the range of feature (F1)	44
3.6	Initial values of the boundary parameters for the central and secondary functions	45
3.7	Adaptation of the initial values of the boundary parameters for the central and secondary functions (horizontal adaptation)	46
3.8	Output for horizontal dimension adaptation of the boundary parameters for the central and secondary functions generated in Fig. 3.6	46
3.9	Vertical dimension adaptation	47
3.10	Output for vertical dimension adaptation of the boundary parameters for the central and secondary functions generated Fig. 3.8	48
3.11	Fuzzy model and classification process	50
4.1	Test rig as the tribological system	57
4.2	Example of erosion rate of the metal surface	58
4.3	Hypothetical behavior of the erosion rate during the operation time	58
4.4	Representations of the operation cycle number 2011 based on fuzzy classifier output space	60
4.5	Representations of the operation cycle number 2011 based on State Space	61
4.6	Representations of the operation cycle number 2011 based on Spectrum Space	62
4.7	Evaluation results of the first dataset the fuzzy classifier output space	63
4.8	Evaluation results of the first dataset in the state space	63

4.9 Evaluation results of the first dataset in the spectrum space 64
4.10 Evaluation results of the second dataset the fuzzy classifier output space 64
4.11 Evaluation results of the second dataset in the state space 65
4.12 Evaluation results of the second dataset in the spectrum space 65

List of Tables

1.1	Overview of classification techniques used with fuzzy logic	7
1.2	Overview of fuzzy rules	8
1.3	Overview of fuzzy partition techniques	9
1.4	Overview of fuzzy rule extraction techniques	9
1.5	Overview of criteria for determining the number of fuzzy rules	10
1.6	Overview of fuzzy rule selection techniques	11
2.1	Canonical form of fuzzy rule-based system	26
4.1	Basic information for the benchmark data sets used	51
4.2	Accuracy values of the AFBA approach and the comparative approaches applied to the used benchmark datasets	52
4.3	Average ranking for the comparative approaches for all of the data sets according to Friedman test (highest is best)	52
4.4	Average ranking for the comparative approaches for all the benchmark data sets according to Wilcoxon test (highest is best)	53
4.5	Average ranking for the comparative approaches for each individual data set according to Wilcoxon test(highest is best)	53
4.6	Basic information of the benchmark structure datasets	53
4.7	Features for the PID dataset	54
4.8	Evaluation results for AFBA compared to other approaches using iris dataset and 3-10cv	55
4.9	Evaluation results for AFBA compared to other approaches using iris dataset and 5-10cv	55
4.10	Evaluation results for AFBA compared to other approaches using iris dataset and 10-10cv	55
4.11	Evaluation results for AFBA compared to other approaches using PID/Pima dataset and 3-10cv	55
4.12	Evaluation results for AFBA compared to other approaches using PID/Pima dataset and 5-10cv	55
4.13	Evaluation results for AFBA compared to other approaches using PID/Pima dataset and 10-10cv	55
4.14	Evaluation results for AFBA compared to other approaches using sonar dataset and 3-10cv	55
4.15	Evaluation results for AFBA compared to other approaches using sonar dataset and 5-10cv	56
4.16	Evaluation results for AFBA compared to other approaches using sonar dataset and 10-10cv	56
4.17	Overview of the related states of surface conditions/erosion rate the oil lubrication	59

Nomenclature

Constants

Symbol	Parameter
A	Antecedent/partition/If-part of the fuzzy rule
A^r	Decomposition of fuzzy rules
C	Consequent/Then-part of the fuzzy rule
C^r	Aggregation of fuzzy rules
CF	Confidence factor of the considered state included in the fuzzy partition within the range of the considered feature
e	Error signal
e_f	Fuzzy error signal
F	Feature
K	Number of the fuzzy partition used in description of the considered state
K_D	Derivate gain
K_I	Integral gain
K_P	Proportional gain
K_P	Proportional gain
L	Number of the features used in description of the considered state
M	Number of the considered states of the to be modeled system/process
NF	Number of used features
$NFMF$	Number of used fuzzy membership functions/fuzzy partitions
NoA	Number of the antecedents of the jth rule included the mth state
NoS	Number of samples of each state included the range of the considered fuzzy membership function
n	Initial number of variables
p	Size of hybrid state vector/number of feature
$\bar{p}$	Mean value of the positions the samples of the considered state within the range of the considered feature
S	State
SW_{BE}	Beginning of the sliding-window feature extractor
SW_{EN}	End of the sliding-window feature extractor
$TNoS$	Total number of samples included the range of the considered fuzzy membership function
X_f	Fuzzy set
α	Symbolic translation of a label in fuzzy linguistic 2-tuple representation
Δ	Difference
ω_{FR}	Weight factor of the fuzzy rule
ω_S	Weight factor of the considered state included in the fuzzy partition within the range of the considered feature
σ	Standard deviation

Abbreviations

AFBA	Adaptive Fuzzy-Based Approach
ANNs	Artificial Neural Networks
EAs	Evolutionary Algorithms
EEG	ElectroEncephaloGram
EFBM	Embedded Fuzzy-Based Modeling
FL	Fuzzy Logic
FCOV	Fuzzy Classifier Output Vector
GAs	Genetic Algorithms
HOV	Homogeneity-Oriented Vector
HSM	Hybrid State Matrix
HSV	Hybrid State Vector
k-nn	k-nearest neighbor
SLT	Statistical Learning Theory
SVMs	Support Vector Machines
VC	Vapnik-Chervonenkis theory

1 Introduction

During the twentieth century, important developments in the theoretical and applied sciences have achieved. These developments led to an increase of human desire/greed in an attempt to identify and explain the phenomena, whether simple or complex, in their environment in order to be able to control them, to guide them in service, and to improve environmental conditions. Additionally, these developments have been dramatically contributed to the evolution and complexity of systems used by the humans in all areas of their life such as in Figure 1.1 [Kon00, Moe06, KGS07, Boc10].

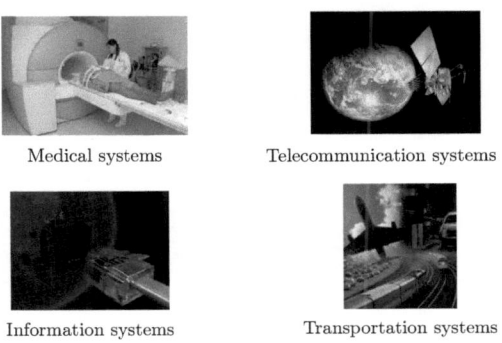

Medical systems   Telecommunication systems

Information systems   Transportation systems

Figure 1.1: Examples of the modern complex systems

However, these developments have presented huge challenges for humans in terms of handling of a large number of information and results from, in the context of determination which of them are beneficial in favor of humans.

The most important challenges are modeling and pattern recognition processes. The modeling is the process of conversion for relevant and/or perceived issues or phenomena to a simpler representation. The mapping typically used in engineering has to be suitable for analysis, interpretation, diagnosis, control, and simulation. The complexity of current systems consisting of interconnected parts is increasing. Additionally, the behavior of these systems includes internal and hidden relationships that are often not directly observable; therefore, the related modeling process becomes more complex and difficult [Kon00, Moe06, KGS07, CKLS07, Boc10].

The pattern recognition is defined as an assignment and classification process of the available information/data into the predefined patterns set (such as in the medical application, the person is sick as pattern 1 or healthy as pattern 2). This pattern is also called *class* or *state*. Usually, this classification process is realized by transferring the original information into a new format, so-called *features* or *attributes*, with high distinguishing ability of these considered states/classes [Kun04, SSD07, TK09].

In order to achieve the modeling and pattern recognition processes of the modern systems, theoretical sciences and techniques of the *Computational Intelligence (CI)* and *Artificial Intelligence (AI)* have been introduced/developed (such as support vector machine (SVM), artificial neural networks (ANNs), etc.) [SS10, Ert11]. Figure 1.2 gives a suggested procedure of CI-based pattern recognition for each of time series data as in the electrocardiogram (ECG) system (right side in Figure 1.2) and structured data as in the Magnetic resonance imaging (MRI) system (left side in Figure 1.2).

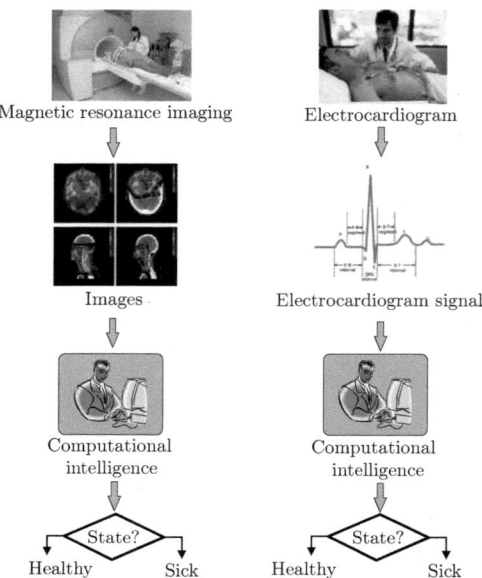

Figure 1.2: Example of procedure of pattern recognition process

The time series signal is defined as sequence of data points, which are measured typically at successive time instants, spaced at uniform time intervals. This signal is converted into a set of statistical characteristics or features to be used by the computational intelligence system for achievement of the assignment and classification tasks [DL07].

The structured data is determined as specific information to be stored based on a methodology of columns and rows. For example, the image shown in the second row of the left side of Figure 1.2 can be converted by means of the image processing technique into the matrix. This matrix contains the columns representing each pixel of this image; while the rows contain the characteristics or features such as the shape area, the center of mass, the moment of inertia, amplitude histogram of each gray-level image, lines, curves, and so on for each considered pixel. Then, this matrix is used by the computational intelligence technique to achieve the assignment and classification process into the sick state or the healthy state [CKLS07, NA08].

For supervised human guidance of complex systems and the realization of complex autonomous systems to replace human direct guidance and human cognition, the main features of human reasoning and perception play a major role in achieving simplification and realizing related cognitive functions and procedures [Wol01]. With the introduction of fuzzy logic [Zad65], simplification and appropriate conclusions based on suitable classification and recognition processes became possible. Thus, fuzzy rule-based systems have been widely applied for classification and recognition goals [Zha08, Fen10, Ros10, WI11].

1.1 State of art

In this section, the state of art of non-fuzzy-based pattern recognition approaches, pattern recognition approaches of time series data, and fuzzy-based pattern recognition approaches will be presented.

1.1.1 Non-fuzzy-based pattern recognition approaches

In the context of this thesis, a "non-fuzzy" term denotes that the design process of recognition approaches is based on the techniques of computational intelligence models (such as support vector machine (SVM), artificial neural networks (ANNs), etc.) except fuzzy model.

The authors of [RAT11] suggested a classification-driven biomedical image retrieval framework as pattern recognition approach. This approach consists of a content-based image retrieval system as feature extraction stage. A relevance feedback technique is used to determine the most useful features set. The selected features set is introduced to a SVM classifier for achieving the desired tasks. Wang and Chen used deterministic learning theory to design a pattern recognition approach for small oscillation faults [WC11]. The authors used ANNs in a training process to obtain knowledge from system dynamics for diagnostic purposes. During the diagnosis phase, a residual set is obtained and average norms are calculated as indicators to monitor potential differences induced by changes in system dynamics. The solution is based on comparison of system-specific properties using signal features measured for error detection. Several packets of wavelet transform and Gabor transform as feature extraction process, ranking and correlation techniques-based feature selection process, and SVM classifier, are used by [TGA+11] to build a texture classification approach of atherosclerosis from the B-mode ultrasound. Boquera et al. developed a recognition process for unconstrained offline handwritten text using optical hybrid hidden Markov and ANN models [BBMM11]. A SVM-based pattern recognition approach with new kernel function, so-called context-dependent, is presented by [SAK11]. A neuromuscular-mechanical fusion-based interface and SVM classifier is developed as pattern recognition approach for locomotion modes and transition modes performed by patients with TF amputations [HZH+11]. An electrooculography (EOG) system as a features extractor of the eye movement and minimum redundancy maximum relevance-based features selector are combined to generate a feature set as input of SVM classifier. This combination constructed a pattern recognition approach in the contribution of [BWGT11]. The contribution of [CLZY11] suggested a pattern recognition approach composed of a 3-D watershed transform to extract a features set of computed tomographic colonography images, then a genetic algorithms (GAs) used as a features selector. The informative features from the GAs stage were used as input of SVM classifier. A bag-of-features technique is suggested by Dardas and Georganas as a features extractor and then a SVM model as a classifier used to detect the patterns of hand gestures [DG11]. The contribution of [TL11] focused on a design of novel classification method based on a probabilistic distance SVM to characterize the sound signals. In this contribution a unification process between structural large margin machine and Laplacian SVM technique in terms of a framework of structural granularity concept is suggested to design the pattern recognition approach. The SVM classifier is used by [MZ11] to propose a novel learning-based unmixing-to-classification conversion model to treat an abundance quantification task. k-nearest neighbor, decision stumps, and linear SVM techniques are integrated to build a real-time textual query-based personal photo retrieval system [LXTL11]. In [LYL+11], a discrete representation of speech signals as feature extraction stage and SVM classifier are used to design the speaker verification system.

Wade et al. suggested a spiking neural network, a Bienenstock-Cooper-Munro learning rule, and spike timing dependent plasticity to design a pattern recognition approach for several classification applications [WMSS10]. A suggested pattern recognition approach of [FKZ10] is a nonlinear approach based on a mixing of linear SVM classifiers. A gene classification approach based on a relative synonymous usage frequency of a codon in feature extraction stage and SVM classifier is proposed by [JMNR09]. In another study, neural networks and Bayesian classifiers were hybridized to design a computational system for automatic tissue identification in wound images. A mean shift procedure and a region-growing strategy were integrated in an effective region segmentation process to generate color and texture features. This features set was then used to train a multilayer perceptron network and Bayesian classifiers for the classification stage [VMM10].

In terms of pattern recognition approaches of unbalanced data sets, the authors of [TZCK09] proposed a novel granular SVM-repetitive understanding algorithm. The contribution of [SGT+09] proposed a new neural network model, so-called graph neural network model, for processing the data used in graph domains. Mohamad and co-workers suggested a pattern recognition approach that depends on a combination of three homogeneous hidden Markov model-based classifiers for recognition of Arabic handwriting [MSM09]. A pattern recognition approach of each static-, dynamic-, and mixed-eccentricity faults of permanent-magnet synchronous motors was designed by [EFR09]. For image recognition problems, a new specialized tiny neural networks-hardware architecture is suggested by [MASR09].

A classification approach of electrocardiogram signals is proposed by [MB08]. In this approach, a particle swarm optimization technique is used to optimize the parameters of SVM classifier. A SVM-based classification scheme for myoelectric control applied to upper limb is suggested by [OH08]. This contribution focused on a usage of segmentation process to generate an initial features set, then an entropy feature technique as a features selector, and SVM technique are used. A self-organization artificial neural network was proposed to construct a pattern recognition approach in an area of image processing [MP08b]. In another study, a wavelet-based neural network was constructed using time–frequency localization functions from wavelet transforms to those from a neural network [MP08a]. To recognize patterns of scoliosis spinal deformity, a total curvature analysis is used to generate features of central axis curve of the spinal deformity patterns in 3-D space [Lin08]. These features are used as input of a multilayer feed-forward artificial neural network trainable based on back-propagation algorithm.

A multiclass SVM classifier of an electroencephalogram signal is suggested by [GÜ07]. This approach consists of a wavelet transform and Lyapunov exponents used in a feature extraction stage and multiclass SVM technique used in the classification stage. A distinguishing process between pathological and non-pathological heart sounds had been achieved based on a pattern recognition approach, which is composed of an automated artificial neural network as classification technique and a combination of a direct ratio and wavelet analysis techniques as the features extractor [dVB07]. In the [DBM07], a pattern recognition approach is based on logistic model tree (LMT) extracting from ANNs. An electromyography signal classifier, so-called cascaded kernel learning machine (CKLM), is built by [LHW07]. The CKLM classifier used an autoregressive model and electromyographic histogram in a feature extraction phase and a combination of one-against-one SVM and generalized discriminant analysis algorithm techniques in a classification phase. A weighted Mahalanobis distance kernel as new function is suggested to be used in the SVM model of classification [WSYT07]. A supervised pattern recognition approach based on an artificial immune classifier was introduced by [ZZGL07] for the remote-sensing imagery application.

A wavelet transform and an ANNs model were incorporated in [SSB06] to build fault detection and classification technique of transmission lines. This wavelet transform was used as a detection module of current waveform analysis in time and wavelet domains.

1.1.2 Pattern recognition approaches of time series data

The authors of [PH11] suggested pattern recognition approach of emotion elicitation procedures in an electroencephalogram. In this approach, higher order crossings and cross-correlation techniques are used to construct feature vector set, which will be later used as input of SVM classifier for achieving the classification tasks. Huang et al. proposed a chaos synchronization-based detector as pattern recognition approach for power-quality disturbances classification in a power system. In the terms of this approach, dynamic error equations technique as feature extraction process and probabilistic neural network as an adaptive classifier were integrated [HLK11]. A sliding bandpass filtering-based time-frequency analysis technique is suggested to extract the features in the Fourier domain [STS11]. A chaos theory-based feature extraction technique was advised by [CPW+11] to generate feature vector of nonlinear dynamical in EEG signal. Then a combination of support feature machine and network-based support vector machine models/classifies are applied to complete the classification process. A nonlinear analysis technique based on a conventional nonlinear statistics was proposed as a core of feature extraction process in [LLL+11] to build features set. The conventional nonlinear statistics included largest Lyapunov exponent and correlation dimension, recurrence and fractal-scaling analysis, and different estimations of entropy. This extracted features set is used as input of classification technique based on a combination of Gaussian mixture models, support vector machine, and fusing generative and discriminative classifiers. A pattern recognition approach of nuclear power plants is proposed by [JGS+11]. The authors of this contribution used symbolic dynamic filtering technique in a feature extraction process and k-nn, linear discriminant analysis, and least squares algorithm in classification process.

A hybrid adaptive filtering technique was suggested as an extraction process of emotion-related electroencephalogram features. Then, higher order crossings classifier was designed for achieving the classification process by using the extracted features set [PH10a]. A discrete wavelet transform derived contexture was integrated with grey-level co-occurrence matrices to construct a feature extraction process of high-resolution satellite imagery. Then in next step, these features were inputted to a maximum-likelihood classifier [OTS10]. A combination of least-squares parameter estimator and autoregressive modeling was advised as the feature extraction process of electroencephalogram signals. These extracted features are used as input SVM-based classification stage [CMP+10]. A higher order crossings analysis technique was employed as a feature extraction process. Then, a HOC-emotion classifier was integrated with the techniques of quadratic discriminant analysis, k-nearest neighbor, Mahalanobis distance, and support vector machines to design electroencephalogram-based emotion recognition approach [PH10b]. A wavelet transform-based feature extraction technique and ANNs model technique as classifier were merged to construct a pattern recognition of faults of electric power system. In this approach, the ANNs model is trained by using particle swarm optimization technique [UGSR10]. A signal turns count of standard deviation parameters of stride interval as a feature extractor and linear discriminant analysis and least squares support vector machine were notified as the pattern recognition approach. The signal turns count was constructed by using Parzen window method in a gait rhythm time series [WK10]. Zhang et al. suggested a high-resolution time-frequency analysis algorithm and matching pursuit techniques to extract a corresponding time-frequency feature description. Then, a principle component analysis technique was used to reduce a size of this description. This new description was inputted to a density-guided k-means clustering classifier [ZLH10]. An electroencephalogram signal was filtered by means of a spatial filtering technique in gamma frequency band to extract informative features to be suitable for a radial basis function classifier-based classification stage [KS10]. A suggested incorporating process between a minimum-redundancy maximum-relevancy (MRMR) filter and a support vector machine recursive feature elimination (SVM-RFE) classifier was achieved by [MR10] as a core of a pattern recognition approach of gene series.

High-voltage signals were analyzed through Fourier and wavelet transforms to extract statistical features in each of the time and wavelet domains as well as spectrum intervals and total energy in the frequency domain. These extracted features were used as input of each of Fisher and Karhunen-Loeve models as the linear classification technique, also as input for the ANNs as non-linear classification technique, for monitoring insulator pollution [BLDC09]. A neural-time-series-prediction-preprocessing technique was a basic of a feature extraction in a proposed pattern recognition approach of electroencephalogram-based brain-computer interface system [Coy09]. A local regression-based stimulus-evoked activity estimation technique was recommended by [WMLL09] as a feature extraction process of local field potential data and SVM classifier. Statistical methods and signal analysis techniques as feature extraction process and SVM classifier were combined to build a recognition approach in [KMN09]. Wavelet decomposition as a filtering technique, nonlinear pulse detection technique as a feature extractor, and statistical pulse as a modeling technique were unified to create a pattern recognition approach of esophageal manometry signal [NDS+09]. Filter banks technique was suggested as a core of features extractor of a material based on image patch exemplars, where statistical characters of extracted features represented an input of nearest neighbor model in classification stage [VZ09].

Common spatial patterns and discriminative spatial patterns techniques were combined to extract relative features of electroencephalogram of brain-computer interface system. A SVM model was later used to classify considered states of signal based on their extracted features [LYWL07]. In contribution of [SWSC07], the radial basis function neural network was proposes as a basic of pattern recognition system of electroencephalogram signal. In this approach, Fourier and wavelet transforms were implemented as a feature extractor of raw electroencephalogram signal. The correlation filters and palm-specific segmentation techniques were feature extractor in palm print classification system [YKS07].

A bin-based distribution model and Gaussian mixture model-based distribution model were used in [PJL+06] to represent time series signal and then these models were input of classification stage consisted of Bayes maximum likelihood and ANNs classifiers.

1.1.3 Fuzzy-based pattern recognition approaches

This subsection summarizes the state of art of the fuzzy-based pattern recognition approaches in the following aspects:

- classification techniques (Table 1.1),
- fuzzy rules (Section 1.2.1),
- fuzzy partition techniques (Section 1.2.2),
- fuzzy rule extraction techniques (Section 1.2.3),
- criteria for determining the number of fuzzy rules (Section 1.2.4), and
- fuzzy rule selection techniques (Section 1.2.5).

1.2 Motivation

Despite the learning and adaptation abilities and the general advantages of approaches such as SVM and ANNs, they are often characterized by an inability to interpret physically their internal relations and parameters, known as the "black box" problem. Acquisition of internal information is often difficult and complex. On the one hand, theoretical or axiomatic approaches used for modeling do not necessarily involve relevant and task-related physical variables. In addition, relations are built based on theoretical assumptions. Well-known techniques have worked well for decades, but these are strongly related to non-complex relations. Related theoretical techniques are able to extend underlying fuzzy-based approaches. More complex phenomena that cross domain borders can often not be modeled easily; in these cases, non-fuzzy approaches based on computational intelligence models (such as SVM and ANNs) often fail. On the

1.2 Motivation

Technique	Related work
Fuzzy logic & Artificial Neural Networks	[IN96, INT99, Ass07, SAC07, NB07, Now08, TNQ08, Wan08a, Wan08b, NB09, CPM09, LWG10, QLT10, SMW+11]
Fuzzy logic	[INT92, INYT94a, INYT94b, NIT94, INT95, INYT95, IMT95, NTI97, INT97, IN97, NNI02, IY05, KA06, BPB+06, WH06, INK06, NYSI06, KK07, Ish07, AFAH07, YLM+07, KN07, SNYI07, AZ08, TNQ08, AMG08, MZK08, IKN08a, IKN08b, HH09, NI09a, NNI09, NI09b, MSR09, LFW09, MZK09, ANHI09, NNI10, NKI10, SEAJK10, SN10, NMI10, INN10, LCG11, AFAH11, IN11, NNI11]
Fuzzy logic & Artificial Neural Networks & Baysian	[AS06, HHE+06, MSG07, JC07, GLQ09]
Fuzzy logic & Support Vector Machines	[LC07, JCS07, ZG07, MTA+08, CL08, BBC10]
Fuzzy logic & Evolutionary Algorithms	[NIT98, CYAP07, PCM09]
Fuzzy logic & k-nearest neighbor	[PP09]

Table 1.1: Overview of classification techniques used with fuzzy logic

other hand, experimental modeling approaches, including pattern recognition methods that yield input–output relations, do not include comprehensive internal structures and explanations for an understanding of internal relations, features or meaningful rules, or do not allow logically related predictions. The comprehensive understanding required can often be achieved by using fuzzy-based pattern recognition approaches because of their phenomenological qualitative modeling basis afforded by fuzzy logic. Here a specific and important modeling task is realized by introduction of a posteriori structuring.

Feature extraction techniques for time series data are introduced in Section 1.1.2. These techniques often suffer from drawbacks in the generation of specific feature types. In addition, they involve complexity in achieving adaptation, adjustment or optimization processes because their structures are based on many parameters.

Motivated by the above-mentioned drawbacks, this contribution introduces a filtering technique as a feature extractor for automatic generation of different types of features. The key property is that the feature extraction process can be used to achieve the tasks desired via a simple mechanism.

In terms of the design of fuzzy-based pattern recognition approaches, this contribution introduces a number of new aspects, as described in the following sections.

1.2.1 Representation form for fuzzy rules

Fuzzy systems are usually built by using conditional statements with linguistic variables to be represented by fuzzy sets and their logical connections for antecedents and consequences. The conditional statement is usually described as follows: *If antecedent, Then consequences*. The antecedent can be understood as a condition that has to be satisfied. A canonical approach is used to build fuzzy systems, whereby the rules chosen are the simplest and most significant ones without loss of generality for the system. The fuzzy rules can take one of the forms listed in Table 1.2, which are the most widely used in fuzzy-based systems. In general, each state considered can be represented by a set of one or more rules.

For the first representation form, the rules are assigned different weight factor ω_{FR} parameter to describe the state. Therefore, a weighting process is required to determine redundant and less important rules within the set. The redundancy or importance of each suggested rule can be determined according to the value of ω_{FR} in relation to the predefined hypothesis space. This weighting process makes the fuzzy inference mechanism more robust because more unreliable rules are excluded and the reliable data-based rules are retained. However, the usage of this form can lead to a loss of information about the other states to be included in the antecedent for the rule set. This information loss means that samples for other states located within the rule range will be assigned to the state described by this set. For the second representation form, in addition to information loss, a weighting process can not be used because of a lack of adjustment parameters. This leads to difficulty in adjusting the set size generated, namely selection of the most important rules, and thus the confidence and robustness of the decision-making process will be reduced. The third form is similar to the second one in terms of drawbacks, but it introduces more flexibility in the interpretation process because of use of the symbolic translation parameter α. The representation form proposed here avoids the above-mentioned drawbacks because of a weighting process that can be realized in both parts of each rule.

Representation form	Related work
If x_{f1} is A^1 ... and x_{fn} is A^{NFMF} then y is C^r with ω_{FR}	[INT92, INYT94a, INYT94b, NIT94, INT95, INYT95, IMT95, NTI97, INT97, IN97, NIT98, INT99, NNI02, IY05, KA06, INK06, NYSI06, Ish07, SNYI07, MZK08, Wan08a, Wan08b, IKN08a, IKN08b, NB09, PCM09, MZK09, NI09a, NNI09, NI09b, ANHI09, NKI10, QLT10, LWG10, SN10, NNI10, INN10, SMW$^+$11, NMI10, IN11, NNI11]
If x_{f1} is A^1 ... and x_{fn} is A^{NFMF} then y is C^r	[BPB$^+$06, CYAP07, SAC07, YLM$^+$07, MTA$^+$08, TNQ08, AMG08, CPM09, MSR09, LFW09, SEAJK10, LCG11]
If x_{f1} is $\left(A^1, \alpha_1\right)$ ... and x_{fn} is $\left(A^{NFMF}, \alpha_{NFMF}\right)$ then y is C^r	[WH06, AFAH07, AFAH11]

Table 1.2: Overview of fuzzy rules

1.2.2 Fuzzy partition technique

An important step in the design of fuzzy systems is the fuzzification or fuzzy partition, which transforms the original crisp data space into a fuzzy set space. Fuzzy partition should achieve a balance between the level of performance required for the system and the complexity associated with the number of fuzzy partitions. If a fuzzy system with less complexity is desired, then the number of fuzzy partitions should be small; this type of fuzzy partition is called coarse partition. Conversely, if the objective is a fuzzy system with best performance regardless of the complexity, then the number of fuzzy partitions should be large; this type of fuzzy partition is called fine partition. Therefore, a fuzzy partition process that is more accurate, adjustable and reliable yields a fuzzy system that is more robust with greater confidence.

These techniques (Table 1.3) have either a lack of precision or a large complexity because of multiple parameters and iterations. Therefore, in this work a statistical characteristics-based technique for fuzzy partition is developed to overcome these problems. The statistical properties of the data are used to determine whether fuzzy partition should be fine or coarse. Thus, in terms of the boundary number and values, the fuzzy partitions are automatically related to the statistical properties of the data. Consequently, the balance between performance level and complexity depends on the intrinsic statistical properties of the data. In contrast to other techniques, no predefined external properties are needed besides signal classification.

1.2 Motivation

Technique	Related work
Human expertise	[KA06, BPB+06, SNYI07, YLM+07, AFAH07, AMG08, Wan08a, Wan08b, MSR09, LFW09, LWG10, SN10, SMW+11, AFAH11]
Clustering	[SAC07, MTA+08, LCG11]
Quantization process	[QLT10]
Decision tree	[SEAJK10]
Hierarchical manner	[TNQ08]
Square distributed simple fuzzy grid	[INT92, INYT94a, NIT94, INT95, INYT95, NKI10, IN11]
Rectangular distributed simple fuzzy grid	[INYT94b, IMT95, NTI97, INT97, NIT98, MZK08, MZK09]
"Don't Care" approach and Rectangular distributed simple fuzzy grid	[IN97, INT99, IY05, INK06, NYSI06, Ish07, IKN08a, IKN08b, ANHI09, NI09a, NNI09, NI09b, QLT10, NMI10, INN10]
Information entropy and rectangular distributed simple fuzzy grid	[NNI02]

Table 1.3: Overview of fuzzy partition techniques

1.2.3 Fuzzy rule extraction

After fuzzy partition, the results should be related to the states being considered using one of the representation forms in Table 1.2. This process is called rule extraction, which requires acceptable accuracy and confidence and should be able to handle complexity. The most widely used techniques are listed in Table 1.4.

Technique	Related work
Human expertise	[KA06, BPB+06, YLM+07, Wan08a, Wan08b, LWG10, SMW+11]
Genetic Algorithms	[INT92, INYT94a, NIT94, INYT94b, INT95, INYT95, IMT95, NTI97, IN97, INT97 NIT98, INT99, IY05, INK06, NYSI06, Ish07, AFAH07, MZK08, IKN08a, IKN08b, ANHI09, MZK09, LFW09, FCM09, NI09a, NNI09, NI09b, QLT10, NMI10, INN10, SN10, NKI10, AFAH11, IN11]
Steady-state Genetic Algorithms	[MZK08, MZK09, PCM09]
Extreme learning machine	[SAC07]

Table 1.4: Overview of fuzzy rule extraction techniques

These techniques suffer from computational cost problems because of the iteration process required to identify a set of rules to represent the desired tasks. The principle, whereby the number of fuzzy rules is equal to the number of states considered, is adopted in the terms of this contribution to overcome the above-mentioned drawbacks. According to this principle, fuzzy rules are extracted without any type of iteration.

1.2.4 Criterion for determining the number of fuzzy rules

In determining the number of fuzzy rules, the principle used to identify the initial size of the fuzzy rule base and the relationship between this size and the parameters (e.g. number of neurons or support vectors) should be considered. The keystone of fuzzy rule-based systems is a rule base. The structure

Criterion	Related work
$\binom{NF}{2} \times 2^{NFMF}$	[SN10]
$2^{NF} - 2$	[QLT10]
Number of selected single granularity	[NNI10]
$NFMF^{NF}$	[INT92, INT95, IMT95, NTI97, INT97, IN97, NIT98, INT99, NNI02, IY05, INK06, NYSI06, Ish07, MZK08, IKN08a, IKN08b, NI09a, NNI09, NI09b, ANHI09, NKI10, SEAJK10, NMI10, INN10, IN11, NNI11]
Human expertise	[KA06, YLM[+]07, Wan08a, Wan08b, MSR09, LFW09, LWG10]
Number of neurons	[CPM09, SMW[+]11]
Number of used clusters	[SAC07, MTA[+]08, LCG11]
$1/6 * NFMF * (NFMF + 1) * (2 * NFMF + 1) - 1$	[INT92, INYT94a, INYT94b, INYT95]

Table 1.5: Overview of criteria for determining the number of fuzzy rules

and size of this base, as determined by the number of rules used, have the greatest influence on the performance of the system. The structure of the base is determined by the fuzzy partition and the fuzzy rule extraction processes. In addition, the number of rules is strongly related to the classification technique used. For example, the number of rules is equal to the number of hidden neurons for combinations of ANNs and fuzzy logic (row 7 in Table 1.5) or to the number of clusters for combinations of fuzzy logic and clustering (row 8 in Table 1.5). If the classification approach is based only on fuzzy logic, the size of the base depends on the number of fuzzy partitions ($NFMF$) generated from the fuzzification process (row 9 in Table 1.5), on the number of feature space dimensions (NF, row 3 in Table 1.5), or on both of them (row 2 in Table 1.5). Therefore, the base size is proportional to the number of neurons and clusters; while this size is exponential to the values of $NFMF$ and NF. Consequently, the effects of complexity and dimensionality problems are strongly related to these criteria. The present study introduces a new idea to avoid complexity and dimensionality problems, whereby the number of fuzzy rules and the parameters $NFMF$ and NF are separated when designing a fuzzy rule-based system. This involves restricting the number of fuzzy rules to the number of states considered. Then the number of fuzzy rules is independent from $NFMF$ and NF and from the parameters of the classification technique.

1.2.5 Fuzzy rule selection

An initial rule set is generated according to the fuzzy partition, rule extraction processes, and the criterion for determining the number of fuzzy rules. This set can include both important and redundant rules with respect to the desired tasks. Therefore, it is necessary to separate the rules to yield a rule base with high accuracy and simple interpretation possibilities. This process should be achieved with the lowest computational cost in the shortest time possible. The most important techniques for fuzzy rule selection are shown in Table 1.6.

These techniques (genetic algorithms, matching and heuristic specification methods) generally suffer

Technique	Related work
Human expertise	[AFAH07, Wan08a, Wan08b, LWG10, AFAH11]
Genetic Algorithms	[INYT94a, INYT94b, INT95, INYT95, IMT95, IN97, NIT98, KA06, INK06, AFAH07, Ish07, MTA$^+$08, MZK08, IKN08a, LFW09, PCM09, ANHI09, MZK09, NI09a, NNI09, NI09b, NKI10, QLT10, SN10, NNI10, NMI10, INN10, NNI11, AFAH11]
Matching principle	[TNQ08]
Heuristic specification methods	[IY05]

Table 1.6: Overview of fuzzy rule selection techniques

from large computational costs because of the iteration process involved. To overcome these drawbacks, the suggested approach does not include any rule selection because of the assumption that the number of fuzzy rules has to equal the number of states considered.

1.3 Organization of the work

In Chapter 2, pattern recognition and fuzzy rule-based systems are presented in terms of the definition, structure, concepts, and techniques. The adaptive fuzzy-based approach (AFBA) used within this thesis/work as the pattern recognition system is described in detail in Chapter 3. The application and verification of the suggested approach and comparison of the results with those for other methods using benchmark data are illustrated in Chapter 4. Finally, in Chapter 5, this thesis/work is summarized; the scientific contribution and the future aspects are introduced.

2 Pattern recognition and fuzzy rule-based systems

In this chapter, the concepts of pattern recognition and fuzzy rule-based systems are introduced.

2.1 Pattern recognition systems

In this section, definition, concepts, and most widely used structures of pattern recognition systems will be introduced in detail.

2.1.1 Definition

"pattern recognition is the scientific discipline whose goal is the classification of objects into a number of categories or classes" [TK09].

An objective of the pattern recognition process is to discover regularities and similarities hidden in the considered data by automatic computer-based algorithms. A goal of this discovery process is to accomplish several tasks such as classification process. The classification process can be understood as a process to classify/relate/assign an object/event, so-called *patterns* into/with one of predefined categories, so-called *classes*, *states* or *categories*. The classification is realized by using a set of quantities/measurements, so-called *features* or *attributes* [Abe01, Kun04, HDRT04, Bis06, CM07, Abe10, Ros10].

2.1.2 Basic concepts

The pattern recognition system includes the following basic concepts [HDRT04, Kun04, GGNZ06, TK09, Abe10, Ros10]

1. input space contains objects to be classified,

2. output space contains related states/classes/categories/targets,

3. state/target is understood as a linguistic expression or description distinguishing different present conditions of system/process to be considered,

4. training data set contains the considered objects/events. The training data can be divided into two types

    - *labeled data*, in which the states/classes/categories/targets of considered objects are known as in case of supervised pattern recognition system,

    - *unlabeled data*, in which the states/classes/categories/targets of considered objects are unknown as in case of unsupervised pattern recognition system,

5. feature space contains a set of measurable quantities being able to distinguish the related states/classes/categories/targets,

6. validation data set contains the objects with predefined labels to check the performance of pattern recognition system,

7. test data set contains the unknown objects to be classified,

2.1 Pattern recognition systems

8. generalization ability denotes to available reliability degree for a use of the designed pattern recognition system in evaluation process of any data, which should have the same properties as the training data set, generated from the considered system/process,

9. discriminant/decision function defines boundaries of separation between the predefined states in the feature space, and

10. accuracy/recognition rate indicates an amount of performance and effectiveness of the designed pattern recognition system performing during the evaluation process of unknown data.

2.1.3 Design procedure

The steps of design procedure of the pattern recognition system can be defined by an answer process of the following questions [HDRT04, Kun04, GGNZ06, TK09, Abe10]

1. which method should be used to determine the training data and test data, as well as their size?

2. which method should be used to determine type and quality of the features used for the classification process?

3. what is the size of the used feature space?

4. which method should be used to select the best size of the used feature space?

5. which criteria should be used to determine the best pattern recognition system corresponding to the considered application?

Basic steps in the design process of the supervised pattern recognition system are illustrated in Figure 2.1. As it can be noticed from Figure 2.1, there are feedback arrows between the steps. An existence of these arrows indicates that these steps are not independent from each other. This means that there is a type of interaction between the predefined design hypothesis space for each step during the design process of the approach. This interaction allows achieving optimization/adjustment processes for the current step according to the evaluation process of any previous or later step. This partial optimization/adjustment process contributes to improve the performance of the general pattern recognition system. Additionally, these arrows can indicate that the two or more steps can be integrated into one step such as in the integration process between the feature selection and the classifier design steps [Kun04, SSD07, TK09].

These basic steps are briefly introduced in the next sections.

2.1.4 Data collection

A key step of the design process of supervised pattern recognition system is a collection process of the objects/patterns of the considered system. This step includes the following tasks

1. determination of each training and test data set, if necessary of the validation data set as well, and

2. determination of the targets/labels vector.

The first task is very important because the suitable selection of these data sets and their sizes will lead to high efficiency design process [Kun04, AKS04, SSD07].

The division/determination methods of the data set can be achieved by means of one of the following most widely used methods [Kun04]

- resubstitution (R-method)
  In this method, the available data is used for each of the training and testing phases.

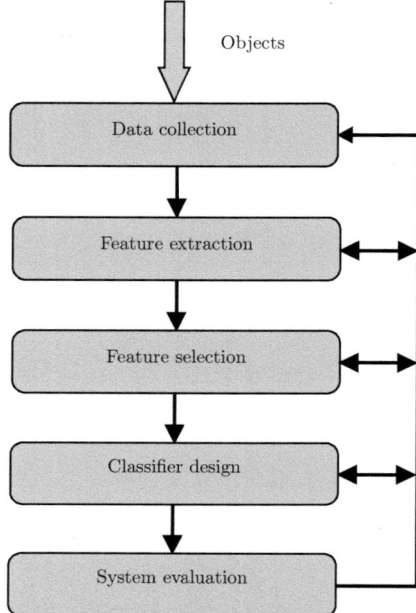

Figure 2.1: Basic steps in the design procedure

- hold-out (H-method)
  In this method, the available data is equally divided into two datasets. The first one is used in the training/learning phase; while the second one is used in the testing phase.

- cross-validation
  In this method, the available data is randomly divided into an integer number of sets. The (integer number-1) sets are used in the training/learning phase and the remaining set is used in the testing phase. For example, 10-fold cross-validation method means the original data will be divided into ten subsets, where nine subsets are for training phase and one is for testing phase.

In the second task, the targets and labels vector can be determined by using a human classification of objects of considered application or automated clustering method [GGNZ06, CKLS07].

2.1.5 Feature extraction

"The purpose of feature extraction is the measurement of those attributes of patterns that are most pertinent to a given classification task" [CKLS07].

The feature extraction can be defined as a trained artificial intelligent subsystem to treat original data for generating a series of semantic quantities, so-called *features*, representing useful information/facts hidden behind these data. These information/facts are used to achieve desired demands of considered application [Moe06]. The authors of [GJN05] defined the feature extraction as the process of construction of identification tools, namely features, to obtain potential relations between the data of the given system/process and their possible classes/states. These potential relations are often hidden and difficult to understand by human perception.

The generalization ability of designed pattern recognition system is strongly related to the type and quality of used features [Abe01, Kun04, GGNZ06, DZ07, CKLS07, TK09]. Hence, the feature extraction should be carefully realized to discover all useful hidden information in the original data [GGNZ06].

Automated feature extraction can be achieved by geometric, structural, and transformation methods [CKLS07, NA08]. The transformation methods are divided into linear methods such as (principle component analysis (PCA), linear discriminant analysis (LDA), independent component analysis (ICA), singular value decomposition (SVD), and nonnegative matrix factorization (NMF)) and nonlinear methods such as (multidimensional scaling (MDS) and artificial neural networks (ANNs)) [CKLS07, LM08, Ize08, TK09, Abe10, MMS11, RACS11].

2.1.6 Feature selection

Feature selection process is introduced/defined as a process of separating relevant from irrelevant features according to the considered classes/states [GGNZ06, CKLS07].

In this context, a term of importance/significance is defined as effect of each suggested feature in the *recognition rate* of the designed pattern recognition system. This effect can be improvement or reduction of the related recognition rate [GGNZ06, CKLS07, Abe10]. From the statistical point of view, the feature selection is a process to define a threshold to distinguish between relevant and irrelevant features based on the predefined hypothesis space [GGNZ06].

Feature selection techniques can be classified according to a interactive relationship between the predefined design hypothesis space of each used classification technique and feature selection process as follows [CKLS07]

1. Filter techniques

    A filter feature selection technique, which also is known as ranking method, is based on a filtering process of original features set via their intrinsic properties and independent of the used classification technique [GGNZ06, CKLS07]. In this technique, a research process is only realized to find out each feature (related significance) to be able to represent crucial aspects (for example state's changes of the considered signal in the time). The filter method is an identification process of the feature's significance/ranking without any optimization process of the designed classifier to fit the predefined hypothesis space [GGNZ06, CKLS07].

    The feature ranking/significance can be determined by different criteria such as Bhattacharya probabilistic distance, fuzzy parameters, recognition rate, mutual information, ellipsoids, and hyperboxes intersection regions among the considered crucial aspects [Abe01, GGNZ06, SIL07, CKLS07].

    Advantages of filter technique are very suitable and useful in the case of the large volumes of data, simple and fast structure of computational algorithm, independent of the type of the classification technique/method [Abe01, GGNZ06, SIL07, CKLS07]. A main drawback is that this technique does not take care of the predefined hypothesis space of the classifier design process during the feature selection process [Abe01, GGNZ06, SIL07, CKLS07].

2. Wrapper techniques

    Wrapper technique introduces integration idea between the predefined hypothesis space of the design process of the classifier and the feature selection process. In this technique, the feature selection process is wrapped around a classification/learning algorithm, as well as taking into account effects of interaction relations between the suggested features to build the classifier. The wrapper method is defined as a synchronous process between an identification process of the feature's significance and the optimization process of the classifier using each suggested feature [GGNZ06, SIL07, CKLS07]. According to a generation principle of the features subset, the wrapper technique can be distinguished as deterministic and randomized techniques [SIL07]. The wrapper technique suffers from high computational cost and overfitting problems [OBC06, GGNZ06, SIL07].

The wrapper technique can be realized by one of the following methods [OBC06, GGNZ06, SIL07]

- sequential pruning method (sequential backwards selection),
- sequential growing method (sequential forward selection),
- backtracking during search method, beam search method, floating search,
- oscillating search method,
- stochastic search,
- decision trees, and
- naive Bayes methods (genetic algorithms and simulated annealing).

3. Hybrid techniques

   This type of feature selection technique is an attempt to fuse all filter and wrapper techniques to reinforce their advantages and to improve/remove their drawbacks as far as possible. The hybrid technique uses firstly one of the filter techniques to reduce the original feature space by defining the most useful feature subsets, then the suitable wrapper technique is used to select the suitable feature subsets building the pattern recognition system satisfying the predefined hypotheses space [GGNZ06, CKLS07].

4. Embedded techniques

   Embedded technique aims to establish the feature selection process inside the design process of the classifier. The design process of the classifier is strongly related to both the feature subsets and the hypotheses spaces. Like with the wrapper methods, the embedded method considers effects of interaction relations among the proposed features. The embedded methods are less computationally intensive compared with the wrapper methods. The embedded technique can be realized by sequential forward method, backward elimination method, weighted naive Bayes, or nested methods [SIL07].

Objective functions of the feature selection techniques can be minimum error (ME), minimum subset (MS), or multicriteria compromise (CP). These functions are generally related to the wrapper, hybrid, and embedded techniques [CKLS07].

The feature selection techniques often need the following evaluation criteria such as

1. data intrinsic measures including the following measurements such as distance, information, and consistency.

2. performance of learning algorithms including the minimum subset (MS) and multicriteria compromise (CP) objectives.

2.1.7 Classifier design

Basically, the supervised pattern recognition system includes the original data and state spaces. The design process of pattern recognition system purposes to relate/map these spaces to classify correctly any unknown samples into the targets/label/state as far as possible. The relation/mapping process is realized using tool/mechanism, so-called *classifier* or *classification method*. The classifier is built based on mathematical and/or computing techniques [Abe01, GGNZ06, CKLS07, Abe10].

2.1 Pattern recognition systems

Categorisation principles of classifier

The classifiers can be categorized into *parametric* or *nonparametric* [Moe06, TK09, Abe10, MD11]. In the parametric classifier, the kind of design/development approach of classifier is based on a prior knowledge about the data distribution as in the case of Gaussian classifier. The classifier, wherein the design/development approach does not need any prior knowledge about the data distribution, is nonparametric such as in the case of neural network, fuzzy, and SVM classifiers. According to the type of the discriminant/decision function used in separating the state space into subspaces of considered states, the classifiers can be distinguished to

- *linear* classifier
  In this type, the discriminant/decision function is linear, and

- *nonlinear* classifier
  In this type, the discriminant/decision function is nonlinear [Moe06, TK09, Abe10, MD11].

Classification methods/classifier types

In this sequel, a review of most important methods used in the design process of classifier is given.

Artificial Neural Networks (ANNs) classifier

"A neural network is a massively parallel distributed processor that has a natural propensity for storing experimental knowledge and making it available for use" [Aiz11].

Also a neural network is understood as a process of information processing in context of a structure based on a structure inspired by the way biological nervous systems, such as the brain, process information [YKTZ11]. In the terms of modeling process, the neural networks can be classified as complementary method of the classical modeling methods or alternative methods. The neural networks provide the high efficient modeling method of the systems, in which there is availability of sufficient data to link considered input-output spaces, especially in the case of systems that require a real time application [AKS04]. In context of the pattern recognition process, the neural networks are understood as a flexible, adaptive learning system with analysis ability of observed data generating from the considered system/process to discover the patterns, which are used to develop nonlinear system models, including a reliable prediction ability and promising solution for many real-world problems [Sam07]. A learning process of ANNs classifier can be achieved by using the learning/training samples/data containing related information about the considered environment. The structure of ANNs is able to acquire knowledge contained in the training data. The obtainable knowledge, which is saved in their synaptic weights, is used to realize the model of the considered system/process [AKS04, Sam07, YKTZ11].

A basic unit of the neural network is a *neuron*, in short (N), consisting of the following elements as illustrated in Figure 2.2.

According to [AKS04, Sam07, YKTZ11], a functional procedure of the neuron is described as follows

Inputs of the neuron are n input signals/features vector $x_1, x_2, ..., x_n$ coming from the considered system/environment. Then $n+1$ weights $\omega_0, \omega_1, \omega_2, ..., \omega_n$ are used to construct a corresponding weighting vector, where the first weight ω_0 is called *bias* or *free weight* because there is not any input signal connected with it. These input and weighting vectors are received, accumulated, or summed by a summing function to produce an output Z based on the equation $Z = \omega_0 + \omega_1 x_1 + \omega_2 x_2 + ... + \omega_n x_n$. The Z output is calculated based on an activation function φ to generate a neuron output. The mapping

18   Chapter. 2   Pattern recognition and fuzzy rule-based systems

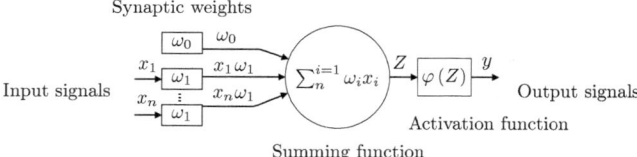

Figure 2.2: Neuron model

function between the input and output spaces in the neuron is defined as follows

$$y = f(x_1, x_2, ..., x_n)) = \varphi(Z) = \varphi(\omega_0 + \omega_1 x_1 + \omega_2 x_2 + ... + \omega_n x_n). \tag{2.1}$$

Most used types of the activation function φ are illustrated in Figure 2.3 [Abe01, AKS04, Sam07, YKTZ11].

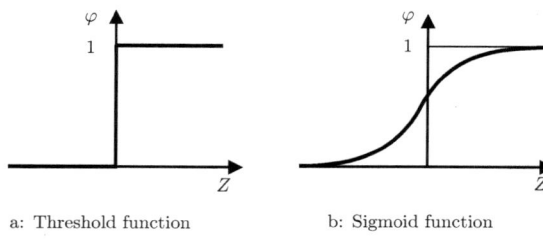

a: Threshold function   b: Sigmoid function

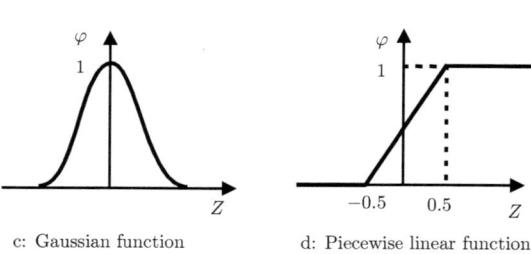

c: Gaussian function   d: Piecewise linear function

Figure 2.3: Activation functions

The most common type of the neural networks classifiers is multilayer neural network classifier (Figure 2.4). The structure of classifier consists of three layers

- *input* layer
  This layer includes input units to spread the input data coming from the surrounding environmental, which can be the outside world or other neural networks, to units of next layer,

2.1 Pattern recognition systems

- *hidden* layer
  This layer is an interface layer between input-output neural network for accumulating and processing the input data. Its name is based on the fact that this layer is related indirectly with the surrounding environmental, and

- *output* layer
  A task of this layer is to offer the information processed in the hidden layers to the surrounding environmental [Abe01, AKS04, Sam07, YKTZ11, Aiz11].

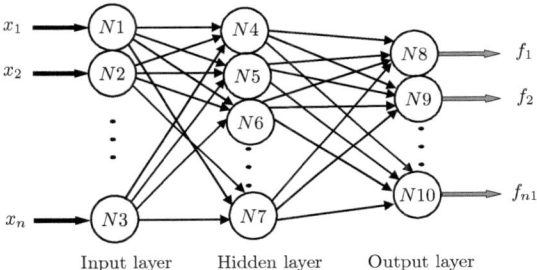

Figure 2.4: Multilayer neural network

The structure of neural network should be activated by the learning process based on a repeated exposure to training data to achieve an incremental change of the connection strengths (synaptic weights) until the network produces a correct output [Abe01, AKS04, Sam07, YCSN10, YKTZ11, Aiz11]

Support Vector Machine (SVM) classifier

Support vector machines (SVMs) are learning methods based on supervised technique for generation of input-output mapping functions dependent of a set of labeled training data. The task of the generated mapping function is to categorize the input space into the considered states [OD08]. According to [WK09, MD11] a support vector machine (SVM) is defined as a binary classifier to abstract a suitable decision boundary in multi-dimensional input/feature space based on an appropriate sub-set, so-called *support vectors*, of the training set of vectors. From another viewpoint, the SVM is a machine learning technique based on a mixture between linear modeling and examples-based learning to select a small number of critical boundary examples, which are called *support vectors*, for each considered class/state within the input/feature space to build a new function distinguishing the two classes/states with biggest range/margin as far as possible [HDRT04, WF05, WM08]. In terms of the statistical learning theory (SLT)/Vapnik-Chervonenkis (VC) theory, the SVM is a systematic and practical application/tool including a theoretical framework of a predictive learning idea to reduce/minimize the classification errors rate generated between usage of the classifier with the training data and its usage with the test data [Vap95, Vap98]. The standpoint of [Bis06, CKLS07] explains that the support vector machine introduces a suitable tool/approach to discover the solution, which includes smallest generalization error for the classification process of the input/feature space, among all the available solutions. This suitable tool/approach is based on a concept of *margin*, which is understood as smallest distance between the suggested decision boundary and any sample in the input/feature space. The support vector machines discover the maximum decision boundary among all available boundaries in the input/feature space. The maximum margin solution/boundary can be determined based on *statistical learning theory*, also known as *computational learning theory*, therefore, the SVM is the maximum margin classifier.

The decision function, in short D, of the SVM classifier for the linearly separable training inputs $(x_1, x_2, ..., x_i, ..., x_n)$ belong to S_1 and S_2 with labels $y_i = 1$ for S_1 and $y_i = -1$ for S_2 is defined as follows [Abe01, OD08, HAM09, WK09, Abe10, MD11]

$$D(x) = \omega^T x + b, \tag{2.2}$$

where ω is weight vector, b is a bias/threshold. In terms of the two labels this decision function is defined by

$$D(x) = \omega^T x + b \begin{cases} > 0 & \text{for } y_i = 1, \\ < 0 & \text{for } y_i = -1. \end{cases} \tag{2.3}$$

In case of the linearly separable training inputs and in terms of the two labels this decision function is defined by

$$D(x) = \omega^T x + b \begin{cases} \geq 1 & \text{for } y_i = 1, \\ \leq -1 & \text{for } y_i = -1. \end{cases} \tag{2.4}$$

The equation 2.4 is equivalent to

$$y_i \left(\omega^T x + b \right) \geq 1 \text{ for } i = 1, 2, ..., n. \tag{2.5}$$

Th separating hyperplane of the n dimensional input/feature space is calculated as follows

$$D(x) = \omega^T x + b = c \text{ for } -1 < c < 1. \tag{2.6}$$

The equations (2.2) to (2.6) are graphically illustrated in in Figure 2.5.

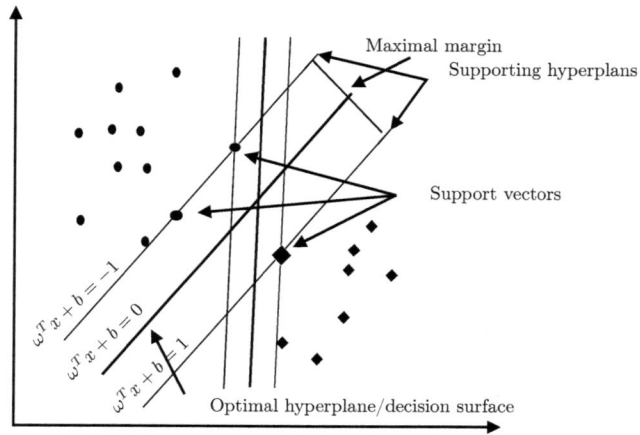

Figure 2.5: Optimal hyperplane and maximal margin in feature space

Nearest neighbor classifier

All training data are used in nearest-neighbor classifiers to build templates of the classification process for a given input vector. The classification process is based on discovery procedure for the template to be nearest to the considered input sample. Then state of this template is state of the considered input sample. In general, based on clustering process, the input/feature space is divided in k templates in k-nearest-neighbor (k-nn) classifier. Then based on distance measurements, such as Euclidean, a distance between each considered input sample and center of each template is calculated to determine the corresponding state dependent of the calculated minimum distance [Bis06, Abe10, MD11].

2.1.8 System evaluation

In this step, the performance and effectiveness of the designed classifier is evaluated based on a calculation of the *recognition rate* or *recognition accuracy*, which is defined as the percentage ratio between the number of the samples with correct classification and the total number of the samples, to determine whether this classifier fits the predefined hypothesis space [TK09, Abe10].

2.2 Fuzzy rule-based systems

In this section, the definition, concepts, fuzzy logic as well as their application in the fields of pattern recognition and control are described.

2.2.1 Basic concepts

In this section, the concepts of fuzzy logic, fuzzy set, fuzzy membership function, fuzzy operation, and fuzzy rule-based system are introduced.

Fuzzy logic

Fuzzy logic introduced by [Zad65] consists of the concepts of the classical logic and the Lukasiewicz sets, namely multiple-valued logic, represented by a new definition of gradual membership idea.

This logic is based on the following assumptions

1. large attention of the gradual interval between *true* and *false* words and also between *white* and *black* colors and

2. mathematics can be able to link between language expressions and human intelligence.

The fuzzy logic provides several linguistic expressions scopes such as

1. scope of *quantification* expressions consisting of *all, most, many, about half, few, no*, represents a gradualness of status at a measurement of some quantity,

2. scope of *usuality* expressions consisting of *always, frequently, often, occasionally, seldom, never*, represents a gradualness of status of event frequency, and

3. scope of *likelihood* expressions consisting of *certain, likely, uncertain, unlikely, certainly no*, explains a gradualness of status of the event probability.

Fuzzy set

Fuzzy set is a set that has elements with degrees of membership (μ) ranging between $[0, 1]$. The fuzzy set, denoted by X_f can be represented by one of the forms

- $X_f = \{(x, \mu_{X_{f1}}(x)) \,|\, x \in X_f\}$.
- $X_f = \mu_1/x_1 + \mu_2/x_2 + \mu_3/x_3 + \ldots + \mu_{NFMF}/x_n$.
- membership function μ_{X_f} (Most used membership functions will be later illustrated in Section 2.2.1).

Here (/) and (+) do not denote division and summation processes, respectively, but they denote a connection of terms of a union of single-term (entity and its membership value) subsets [KB06, SSD07, Ros10, Fen10, SS10, YKTZ11, WI11].

Fuzzy membership function

A fuzzy membership function is defined as a representation form of the fuzzy set. This function is defined as a mapping function between the crisp space and fuzzy space. The conversion/mapping process is called the *fuzzification* process [KB06, SSD07, Ros10, Fen10, SS10, YKTZ11, WI11]. These functions can be represented graphical as in Figure 2.6 or mathematical as follows [KB06, SSD07, Ros10, Fen10, SS10, YKTZ11, WI11]

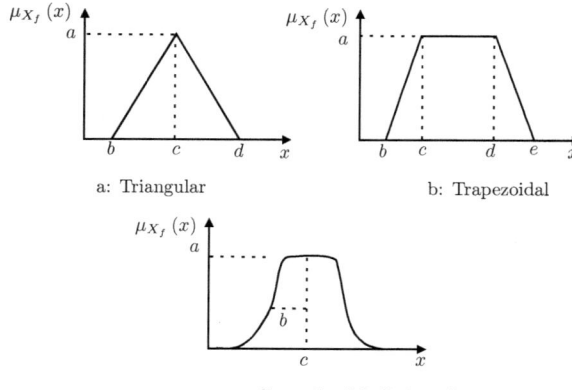

a: Triangular b: Trapezoidal

c: Generalized Bell-shaped

Figure 2.6: Forms of the standard membership functions (triangular, trapezoidal, and generalized Bell-shaped)

- triangular membership function is defined as follows (see Figure 2.6a)

$$\mu_{X_f}(x) = \begin{cases} a(b-x)/(b-c) \,; x \in [b, c] \\ a \,; x = c \\ a(d-x)/(d-c) \,; x \in [c, d] \\ 0 \,; otherwise \end{cases} \quad (2.7)$$

2.2 Fuzzy rule-based systems

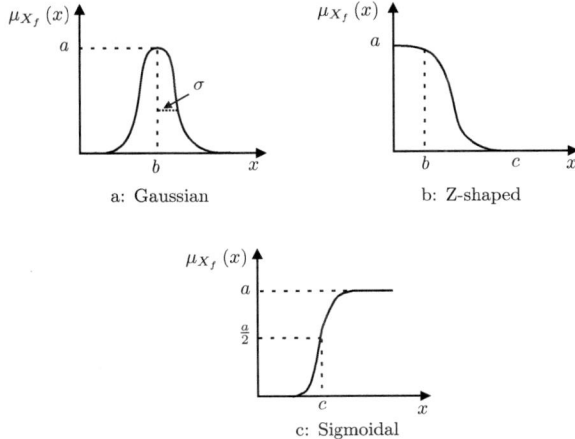

Figure 2.7: Forms of the standard membership functions (Gaussian, Z-shaped, and Sigmoidal)

- trapezoidal membership function is defined as follows (see Figure 2.6b)

$$\mu_{X_f}(x) = \begin{cases} a(b-x)/(b-c) ; x \in [b,c) \\ a ; x \in [c,d] \\ a(e-x)/(e-d) ; x \in (d,e] \\ 0 ; otherwise \end{cases} \quad (2.8)$$

- generalized Bell-shaped built-in membership function (see Figure 2.6c)

$$\mu_{X_f}(x) = \frac{1}{1 + \left|\frac{(x-c)}{a}\right|^{2b}} \quad (2.9)$$

- Gaussian membership function is defined as follows (see Figure 2.7a)

$$\mu_{X_f}(x) = a \, exp \frac{(-x-b)^2}{2\sigma^2} \quad (2.10)$$

- Z-shaped built-in membership function is defined as follows (see Figure 2.7b)

$$\mu_{X_f}(x) = \begin{cases} a ; x \leq b \\ 1 - 2\left(\frac{x-b}{c-b}\right)^2 ; b \leq x \leq \left(\frac{b+c}{2}\right) \\ 2\left(c - \frac{x}{c-b}\right)^2 ; \left(\frac{b+c}{2}\right) \leq x \leq c \\ 0 ; x \geq c \end{cases} \quad (2.11)$$

- sigmoidal membership function is defined as follows (see Figure 2.7c)

$$\mu_{X_f}(x) = \frac{1}{1 + exp(-a(x-c))} \quad (2.12)$$

Most specifications of the membership function are the core, support, and boundary as illustrated in Figure 2.8.

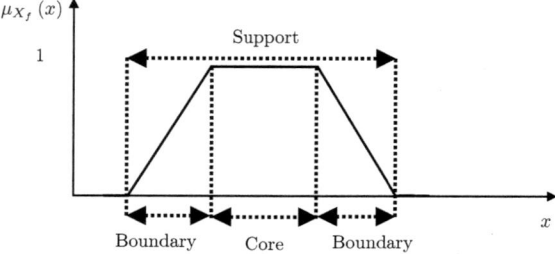

Figure 2.8: Specifications of the membership function

The membership function concept includes the following assumptions

1. all above-mentioned forms of the membership functions have a parameters set of control in the shape (width, height, and slope) of the membership function. In general, a criterion of their definition/determination/adjustment is based either on human experiences and/or knowledge-based about the considered system/process or known data-based generated by the considered system/process, and

2. an interpretation of the membership function does never denote to a *probability* of entity in the input space, but to its *possibility*. Therefore, the sum/integration of the membership functions for whole input space is not necessarily equal to one [CP01, Ibr04, Rut04, EVW05, KB06, SSD07, Ros10, Fen10, SS10, YKTZ11, WI11].

Fuzzy operations

Operations of the fuzzy sets can be distinguished as follows [KB06, SSD07, Ros10, Fen10, SS10, YKTZ11, WI11]

1. logic operations

    - union
    $\mu_{X_{f1} \cup X_{f2}}(x) = \mu_{X_{f1}}(x) \vee \mu_{X_{f2}}(x)$
    - intersection
    $\mu_{X_{f1} \cap X_{f2}}(x) = \mu_{X_{f1}}(x) \wedge \mu_{X_{f2}}(x)$
    - complement
    $\mu_{\overline{X_{f1}}}(x) = 1 - \mu_{X_{f1}}(x)$

2. T-norm, S-norm, and negation operators
   These operators are as follows

    - t-norm
    A t-norm, which is denoted by an operator ($\mathbf{T}$), is defined for the two fuzzy numbers/variables (x_1, x_2) as follows
    $\mathbf{T} : [0,1] \times [0,1] \rightarrow [0,1] =$
    $(x_1 \textbf{ AND } x_2) \equiv (x_1 \textbf{ T } x_2) = \mathbf{T}(x_1, x_2)$

2.2 Fuzzy rule-based systems

- t-conorm
  A t-conorm, which is denoted by an operator (**S**), is defined for the two fuzzy numbers/variables (x_1, x_2) as follows
  $\mathbf{S} : [0, 1] \times [0, 1] \to [0, 1] =$
  $(x_1 \textbf{ OR } x_2) \equiv (x_1 \textbf{ S } x_2) = \mathbf{S}(x_1, x_2)$

- negation
  A negation, which is denoted by an operator (**N**), is defined as follows $\mathbf{N}[0, 1] \to [0, 1] \equiv \mathbf{N}(x) = 1 - x$

These operations can be understood as generalization of (**AND**) and (**OR**) operations, respectively, to deal with the fuzzy numbers.

Properties of Fuzzy set

Important properties of the fuzzy sets can be summarized as [KB06, SSD07, Ros10, Fen10, SS10, YKTZ11, WI11]

- commutativity
  $X_{f1} \cap X_{f2} = X_{f1} \cap X_{f2}$ and $X_{f1} \cup X_{f2} = X_{f1} \cup X_{f2}$.

- associativity
  $X_{f1} \cap (X_{f2} \cap X_{f3}) = (X_{f1} \cap X_{f2}) \cap X_{f3}$ and
  $X_{f1} \cup (X_{f2} \cup X_{f3}) = (X_{f1} \cup X_{f2}) \cup X_{f3}$.

- distributivity
  $X_{f1} \cap (X_{f2} \cup X_{f3}) = (X_{f1} \cap X_{f2}) \cup (X_{f1} \cap X_{f3})$ and
  $X_{f1} \cup (X_{f2} \cap X_{f3}) = (X_{f1} \cup X_{f2}) \cap (X_{f1} \cup X_{f3})$.

- idempotency
  $X_{f1} \cup X_{f1} = X_{f1}$ and $X_{f1} \cap X_{f1} = X_{f1}$.

- identity
  $X_{f1} \cup \varnothing = X_{f1}$, $X_{f1} \cap X_f = X_{f1}$, $X_{f1} \cap \varnothing = \varnothing$, and $X_{f1} \cup X_f = X_f$.

- transitive
  if $X_{f1} \subseteq X_{f2}$ and $X_{f2} \subseteq X_{f3}$, then $X_{f1} \subseteq X_{f3}$.

- involution $\overline{\overline{X_{f1}}} = X_{f1}$.

2.2.2 Fuzzy rule-based system

Definition

Fuzzy systems are usually built using conditional statements with linguistic variables to be represented by fuzzy sets and their logical connections for antecedents and consequences. The conditional statement is usually described as follows

If antecedent, Then consequences.

The antecedent can be understood as a condition that has to be satisfied. A canonical approach is used to build fuzzy systems whereby the rules chosen are the simplest and most significant ones without loss of generality for the system (see Table 2.1) [KB06, Ros10, Fen10, SS10].

Formation of rules

The linguistic variables can be organized by means of the following most widely used types of the canonical formation [KB06, Ros10, Fen10, SS10]

- assignment statements
  In this organization, a combination between the variable and its assigned value is based on usage of an *assignment operator* (=). Therefore this statement is built to put the value of variable into a specific equality regardless of their related conditions.

- conditional statements
  In this organization, a combination between the variable and its assigned value is achieved by realization of some conditions. Therefore, this statement is understood as the fuzzy conditional statement.

Rule 1	IF antecedent A^1 THEN consequence C^1
Rule 2	IF antecedent A^2 THEN consequence C^2
...	...
Rule m	IF antecedent A^m THEN consequence C^m

Table 2.1: Canonical form of fuzzy rule-based system

Decomposition of rules

In general, the fuzzy rule is a compound of the conditions related to many variables and their assigned values. Because of that these variables are fuzzy sets, hence the fuzzy operations and properties can be used to decompose a structure of this fuzzy rule to generate the simple canonical formation. The decomposition process is related to If-part/If antecedent of the fuzzy rule [KB06, Ros10, Fen10, SS10].

The most widely used methods of the decomposition of the rules are as follows [KB06, Ros10, Fen10, SS10]

- multiple conjunction antecedents
  This method is based on a linguistic (AND) connective operator. The structure of the corresponding rule is as follows $If \ x_f \ is \ A^r \ Then \ y \ is \ C^r$,
  where $A^r = A^1 \ AND \ A^2 \ ... \ AND \ A^{NFMF} \equiv A^1 \cap A^2 \ ... \cap A^{NFMF}$.
  The fuzzy membership value of the variable x_f is calculated as follows
  $\mu_{A^r}(x_f) = \min[\mu_{A^1}(x_f), \mu_{A^2}(x_f), ..., \mu_{A^{NFMF}}(x_f)]$.

- multiple disjunctive antecedents
  This method is based on a linguistic (OR) connective operator. The structure of the corresponding rule is as follows $If \ x_f \ is \ A^r \ Then \ y \ is \ C^r$,
  where $A^r = A^1 \ OR \ A^2 \ ... \ OR \ A^{NFMF} \equiv A^1 \cup A^2 \ ... \cup A^{NFMF}$.
  The fuzzy membership value of the variable x_f is calculated as follows
  $\mu_{A^r}(x_f) = \max[\mu_{A^1}(x_f), \mu_{A^2}(x_f), ..., \mu_{A^{NFMF}}(x_f)]$.

Aggregation of rules

The structure of the fuzzy rule-based system include usually many rules. An objective of building this structure is to get one value/conclusion for many values including in these rules to achieve a desired goal such as control, classification, making decision, etc. This objective can be performed in the fuzzy

2.2 Fuzzy rule-based systems

rule-based system based on the *aggregation* process. Therefore, the aggregation process is related to the Then-part/consequences-part of the fuzzy rule [KB06, Ros10, Fen10, SS10].

The aggregation process can be established by using the following most widely used methods [KB06, Ros10, Fen10, SS10]

- conjunctive system of rules
  This method is based on a linguistic (AND) connective operator as follows
  $y = y^1 \ AND \ A^2 \ ... \ AND \ y^{NFMF} \equiv A^1 \wedge A^2 \ ... \ \wedge A^{NFMF}$.
  The fuzzy membership value of the variable y is calculated as follows
  $\mu_{C^r}(y) = \min\left[\mu_{C^1}(y), \mu_{C^2}(y), ..., \mu_{C^{NFMF}}(y)\right]$.

- disjunctive system of rules
  This method is based on a linguistic (OR) connective operator as follows
  $y = y^1 \ OR \ A^2 \ ... \ OR \ y^{NFMF} \equiv A^1 \vee A^2 \ ... \ \vee A^{NFMF}$.
  The fuzzy membership value of the variable y is calculated as follows
  $\mu_{C^r}(y) = \max\left[\mu_{C^1}(y), \mu_{C^2}(y), ..., \mu_{C^{NFMF}}(y)\right]$.

Fuzzy inference system

Fuzzy rule-based system is also known as fuzzy inference system, fuzzy expert system, and fuzzy associative memory. The basic structure of this system is illustrated in Figure 2.9 [KB06, Zha08, SS10, Fen10, Ros10, WI11].

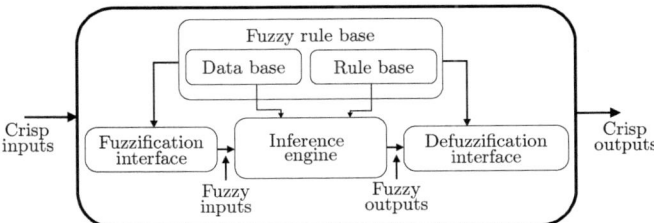

Figure 2.9: Basic structure of the fuzzy-based system

These components/steps are explained as follows

1. fuzzification/fuzzifier component/step
   Fuzzification/fuzzifier component/step can be described as an input of the fuzzy-based system. Task of this component/step is to convert considered crisp real-valued system/process/operations variables/quantities into the normalized fuzzy value/fuzzy set by using one or more the suitable membership function (see Figure 2.6) [Zha08, SS10, Fen10, Ros10, WI11]. According to size, which denotes a number of elements, of support set of used membership functions, the fuzzification/fuzzifier component/step can be one of the two types as follows

   - a singleton fuzzification/fuzzifier component/step
     A used membership function has the support set consisting of one elements/members as in singleton membership function,
   - a non singleton fuzzification/fuzzifier component/step
     A used membership function has the support set consisting of many element/member as in

triangular/ramp membership function, step membership function, trapezoidal membership function, and Gaussian membership function (see Figure 2.6) [Zha08, SS10, Fen10, Ros10, WI11].

The fuzzification process is understood as representation of the system inputs in context of linguistic terms so that rules can be applied in a simple manner to express a complex system [Zha08, SS10, Fen10, Ros10, WI11].

2. knowledge/fuzzy rule base component/step
Knowledge/fuzzy rule base component/step consists of two following bases [Zha08, SS10, Fen10, Ros10, WI11]

   (a) database is used to define each of objects and membership functions used in the fuzzy rules. The database is the declarative part of the knowledge/fuzzy rule base, and

   (b) rule base contains information of how these objects can be used to infer conclusions. The rule base is the procedural part of the knowledge/fuzzy rule base [Zha08, SS10, Fen10, Ros10, WI11].

3. logic inference/inference engine component/step
This component/step is used to perform the inference operations on the suggested fuzzy rules. The inference operations include the following

   - comparison process between the input variables and the membership functions used in the If-part of each rule to generate the membership value of each linguistic label,
   - combination process of the membership values the If-part of each rule by using t-norm to get the *firing strength* of each rule, and
   - generation process of qualified consequents of each rule depending on its firing strength.

4. defuzzification component/step
A prefix (De) in the term of defuzzification indicates a meaning of *do the opposite*, what happens in the fuzzification process. The defuzzification is process to convert a fuzzy output of the inference engine component based on information of rule base into the crisp value that will be sent to execute appropriately the system/operation according to conditions of related input [Zha08, SS10, Fen10, Ros10, WI11].

Defuzzification methods include the following methods [Jam97, CP01, Zha08, SS10, Fen10, Ros10, WI11]

   - centroid method/center of mass/center of gravity
   The defuzzified value/output is calculated as follows

$$x^* = \frac{\int \mu_{X_f}(x) \cdot (x) \, d(x)}{\int \mu_{X_f}(x) \, d(x)} \tag{2.13}$$

   where the symbol ($\int$) denotes algebraic integration.

   - center of largest area
   The center of largest area method can be applied only if the output consists of at least two convex fuzzy subsets, which do not overlap, resulted in a bias toward a side of one membership function.

2.2 Fuzzy rule-based systems

- max-membership/height
  The max-membership/height method can be applied only if the output/function is peaked, where the defuzzified value/output is a value of a domain with the maximum membership value
- weighted average
  The weighted average method can be applied only if the output membership functions are symmetrical, where each membership function is weighted by its maximum membership value. The defuzzified value/output is calculated as follows

$$x^* = \frac{\sum_{i=1}^{n_{mf}} \mu\left(\overline{x_{wi}}\right) \cdot \overline{x_{wi}}}{\sum_{i=1}^{n_{mf}} \mu\left(\overline{x_{wi}}\right)} \qquad (2.14)$$

where the symbol $(\sum)$ denotes algebraic summation and a $(\overline{x_{wi}})$ is a weighted average of the ith fuzzy set.

- mean-max membership/middle-of-maxima
  The defuzzified value/output is calculated as follows

$$x^* = \frac{\sum_{i=1}^{n_{mf}} \overline{x_{wi}}}{n_{mf}} \qquad (2.15)$$

where a $(\overline{x_{wi}})$ is a weighted average of the (i^{th}) fuzzy set.

2.2.3 Application areas of fuzzy rule-based systems

In this section, most important application areas of the fuzzy rule-based systems are introduced.

Control techniques

Control techniques are particularly the most important/earliest application area of the fuzzy rule-based systems. The fuzzy control techniques has shown successful and powerful ability in many practical and industrial applications. Especially in those applications, wherein the related systems can be described as complex nonlinear or even non-analytic. Additionally the fuzzy control introduced an alternative or complementary approach to conventional control techniques in many engineering applications [Zha08, Fen10].

The fuzzy logic control systems are knowledge-based systems, wherein the design process is based on the designer knowledge/experience and/or on available information about the system/problem to define input-output intervals and the describing membership functions. Therefore, the definition process of the input-output intervals and the membership functions can be characterized as *subjective* process [CP01].

The fuzzy logic control approaches can be roughly classified into the following categories [Zha08, SS10, Fen10, Ros10, WI11]

1. conventional fuzzy control or Mamdani fuzzy control
   A basic of Mamdani fuzzy control algorithm is a set of heuristic control rules, and fuzzy sets. The fuzzy logic is used to represent linguistic terms and to evaluate the control rules as follows [SS10, Fen10, Ros10, WI11]
   Rule 1 :
   $If\ x_{f1} = A^{11}\ AND\ If\ x_{f2} = A^{12}\ AND\ ...\ If\ x_{fj} = A^{1j}\ Then\ y = C^1$
   Rule 2 :
   $If\ x_{f1} = A^{21}\ AND\ If\ x_{f2} = A^{22}\ AND\ ...\ If\ x_{fj} = A^{2j}\ Then\ y = C^2$
   ...

Rule m :
If $x_{f1} = A^{m1}$ *AND If* $x_{f2} = A^{m2}$ *AND ... If* $x_{fj} = A^{mj}$ *Then* $y = C^m$.

In the original Mamdani fuzzy control the $t-norm = min$ operator, maximum aggregation method and center of gravity method as defuzzyfication method are used. But later, the variety of the fuzzy operators and the aggregation and defuzzyfication methods are used to develop the Mamdani fuzzy control. The Figure 2.10 illustrates the Mamdani fuzzy control as a 2-input fuzzy rule-based system [SS10, Fen10, Ros10, WI11].

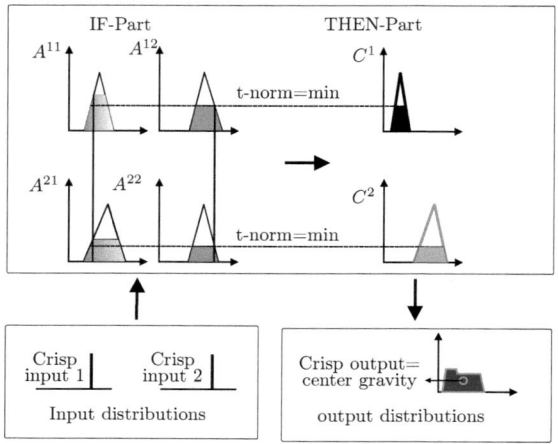

Figure 2.10: Mamdani fuzzy control as the binary fuzzy rule-based system

2. Takagi-Sugeno model-based fuzzy control
A basic of Takagi-Sugeno fuzzy control algorithm is a set of local linear models that are smoothly connected by nonlinear fuzzy rules/membership functions as follows [SS10, Fen10, Ros10, WI11]
Rule 1 :
If $x_{f1} = A^{11}$ *AND If* $x_{f2} = A^{12}$ *AND ... If* $x_{fj} = A^{1j}$ *Then* $y^1 = f_1(x_{f1}, x_{f2}, ..., x_{fi}, x_0)$
Rule 2 :
If $x_{f1} = A^{21}$ *AND If* $x_{f2} = A^{22}$ *AND ... If* $x_{fj} = A^{2j}$ *Then* $y^2 = f_2(x_{f1}, x_{f2}, ..., x_{fi}, , x_0)$
...
Rule m :
If $x_{f1} = A^{m1}$ *AND If* $x_{f2} = A^{m2}$ *AND ... If* $x_{fj} = A^{mj}$ *Then* $y^m = f_m(x_{f1}, x_{f2}, ..., x_{fi}, , x_0)$.
where x_0 is constant value.

The difference between the Mamdani and Takagi-Sugeno model can be distinguished in the Then-part of the used fuzzy rule form. The Mamdani system uses the fuzzy rule with the Then-part based on the fuzzy set, while the Then-part of the fuzzy rule used in the Takagi-Sugeno system is based on a function containing the input variables and constant value.

In the original Takagi-Sugeno fuzzy control the $t-norm = min$ operator, maximum aggregation method and weighted average method as defuzzyfication method are used [SS10, Fen10, Ros10, WI11]. The Figure 2.11 illustrates the Takagi-Sugeno fuzzy control as a 2-input fuzzy rule-based system [SS10, Fen10, Ros10, WI11].

2.2 Fuzzy rule-based systems

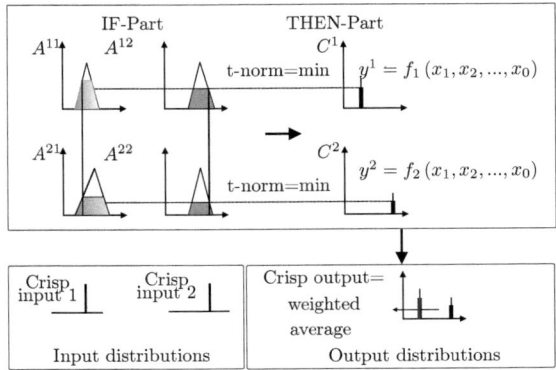

Figure 2.11: Takagi-Sugeno fuzzy control as the binary fuzzy rule-based system

3. fuzzy proportional-integral-derivative (PID) control

   Due to number of distinctive advantages related to PID technique (such as their simple structure, simplicity of design, and low cost of implementation in comparison to many other control methods), the conventional proportional-integral-derivative (PID) controllers are still one of the most widely control techniques adopted in the industry areas for achieving of variety tasks of the control. On the other hand, a performance of the PID controllers is still not satisfactory in the following cases [Zha08, SS10, Fen10, Ros10, WI11]

   - high levels of nonlinearity and uncertainty associated with the considered system and
   - high levels of the control performance specification is very demanding.

The nonlinearity and uncertainty problems can be solved by the fuzzy control using of the fuzzy set theory. Therefore, it is advisable to integrate the advantages each of PID control and fuzzy control to build the control technique to be more robust and adaptive to achieve the control tasks [KB06, Zha08, SS10, Fen10].

The relationship between each input and output of the classical PID controller is (see Figure 2.12a)

$$u = K_P e + K_I e + K_D e \tag{2.16}$$

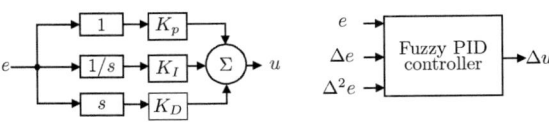

a: Classical PID controller b: Incremental variant

Figure 2.12: Classical PID controller and incremental variant of the fuzzy PID controller

The most widely used integration process of the PID controller and fuzzy logic can be achieved in different variants, e.g. incremental variant (see Figure 2.12b) [KB06, Zha08, SS10, Fen10].

4. neuro-fuzzy control or fuzzy-neuro control
   The neural network control and fuzzy control are of the most popular techniques in the intelligent control areas. There are a number of common features between them such as basically model-free control techniques, storage of knowledge as base to deduce control actions, and robustness property. But their differences can be summarized in distinctive ways of obtaining knowledge, whereas the neuro control acquires knowledge mainly through data training/learning. Either the fuzzy control obtains knowledge via an operator or expert perspective.

   There is a kind of complement between each other in the context of the learning capabilities and high computational efficiency in parallel implementation associated with the neural control and powerful framework for expert knowledge representation provided by fuzzy control. Therefore, a combination/integration process of the neural network control and fuzzy control techniques introduces promising results in generation of better control approaches.

   The neuro-fuzzy control is characterized by no need of existence of information about a mathematical model of the system to be controlled. Thus, this technique offers a new avenue to solve many difficult control problems in real life where the mathematical model of a system might be difficult to obtain [Fen10, WI11].

   Figure 2.13 illustrates a possible structure of the neuro-fuzzy/fuzzy-neuro controller.

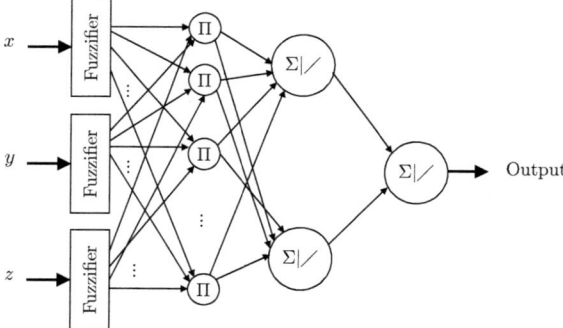

Figure 2.13: Possible structure of a neuro-fuzzy/fuzzy-neuro controller

5. adaptive fuzzy control
   The basic structure of a fuzzy system can be used for implementation of a supervisory/adaptation algorithm. Gain-scheduling techniques are the most frequently used adaptation technique, whereas the operating point and gain coefficients of conventional controller are changed according to the designed nonlinear fuzzy mapping function(s). The fuzzy algorithm can be expressed as an external control element [KB06, Fen10]. Figure 2.14 illustrates the general structure of adaptive fuzzy control systems.

Generally, the above-mentioned classification of the fuzzy logic control approaches is neither unique nor exhaustive and many other different classifications can also be employed. These classifications can be inevitably overlapped to enhance the properties of the designed systems as an addition of adaptive property for the conventional fuzzy control system and for neuro-fuzzy control system, or to tune the fuzzy PID control based on neuro-fuzzy systems [Fen10].

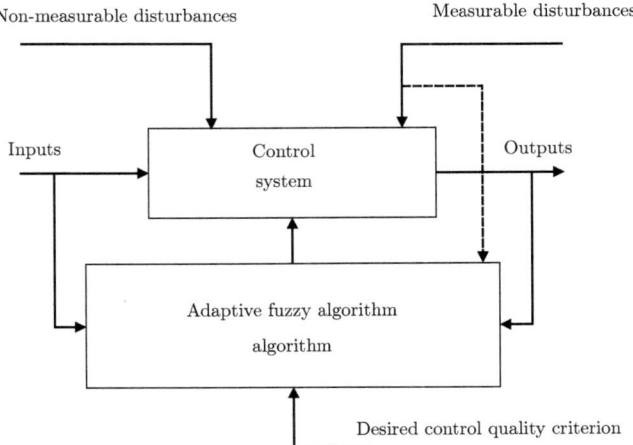

Figure 2.14: General structure of adaptive fuzzy control systems

The assumptions of the design process of fuzzy logic controller are as follows

1. a plant/system/process to be controlled is *observable* and *controllable*.

2. a presence of knowledge/experience/sufficient information about the system in terms of linguistic rules, which could reflect the operational requirements, as well as understanding of engineering for it or a set of data on measurements of input-output, which can be used to extract rules that represent the system,

3. an existence of solutions, which are able to express in term the linguistic rules, can be able to overcome the engineering problems encountered in the system,

4. a control engineer should focus its full attention on getting an *adequate enough* solution of the problem, not an optimum solution, and

5. accuracy/precision of designed controller should be within an acceptable range [Zha08, SS10, Fen10, Ros10, WI11].

The design steps of the fuzzy logic controller can be summarized as follows

- variables (input, states and output) of the considered system should be defined,

- range/interval of each variable is divided into the fuzzy subsets and a linguistic label (i.e. small, large, middle, ect.) and its linguistic hedges (i.e. fast small, very large) of range/interval are determined,

- a suitable membership function, which can be the triangular, trapezoidal, and Gaussian according to a nature of the application/data, is determined for each fuzzy subset in the range/interval of each considered variable,

- the fuzzy relations between the input and/or state fuzzy subsets and the output fuzzy subsets are assigned to create a *rule-base*, namely, If-Then rules,

- a normalization process is achieved to transfer the ranges/intervals of the inputs and outputs variables to [0, 1] range,
- a fuzzification process,
- a fuzzy approximate logic/reasoning is used to infer the output of each rule in the rule-base,
- outputs of each rule in the rule-base is combined in the *aggregation* process,
- a defuzzification is performed to generate a crisp output, which will be used to achieve a certain control purpose [Zha08, SS10, Fen10, Ros10, WI11].

Pattern recognition techniques

The most widely used fuzzy-based pattern recognition systems are explained in the next subsections.

Fuzzy rule-based classifiers/Conventional fuzzy classifier A design process of conventional fuzzy classifier is based on the fuzzy rule-base that is determined according to human experiences/observations on the system/
process. The steps of the conventional classifier design can be summarized [Abe10]

1. ranges of input variables are divided into several non-overlapping intervals,
2. a membership function is defined for each interval,
3. the input space is covered by several non-overlapping hyper rectangles, which represent the decision functions,
4. a fuzzy rule is assigned for each hyper rectangle.

The generalization ability of fuzzy rule-based classifier is based essentially on the number/size of the input variables and the number of the training data, where this ability will be low in case of the small number of training data or in case of a large number of input variables [Abe10].

Fuzzy-neural/neuro-fuzzy classifier (FNNs) Learning and adaptive capacities of traditional forms of fuzzy systems such as fuzzy classifiers can be characterized by a property of *weakness*. But the fuzzy mathematics according to its principles will be able to provide a strong inference mechanism for an approximate reasoning in the terms of cognitive uncertainty. The neural networks can be characterized with the learning, adaptation, generalization, approximation, fault tolerance characters. Additionally, neural networks have strong ability to deal with the features of computational complexity, uncertainty, and nonlinearity characters, which can be escorted to the considered systems. An idea of merging these technologies have been introduced/developed to build *fuzzy neural network* classifier. This classifier introduces the dual ability for avoidance of the individual disadvantages as well as for reinforcement of the individual advantages in the view of a mechanism construction with robust capacity in simulation of many behaviors associated with activities of intelligence and cognition of the human being [GGNZ06]. One type of the fuzzy neural network classifiers can be defined as

"The distributed parallel information processing schemes that employ neuron like processing unit with learning capabilities and fuzzy operations for dealing with fuzzy signals" [GGNZ06].

Most common classes of the (FNNs) are based on two types of configuration

1. mapping of a fuzzy input set to a fuzzy output set by using the fuzzy triangular inputs and outputs,

2.2 Fuzzy rule-based systems

2. mapping of a crisp input set and output signals by using many fuzzy operations and approximate reasoning dependent of the rule-based knowledge framework such as Takagi-Sugeno neuro-fuzzy systems.

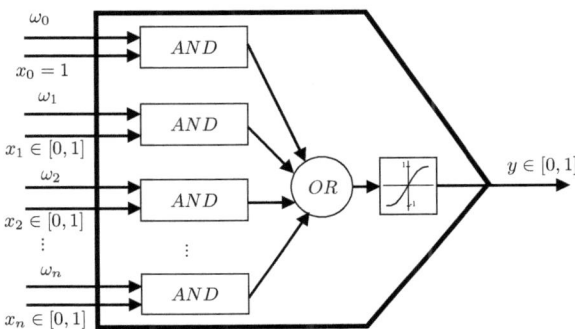

a: OR-AND-type fuzzy neuron

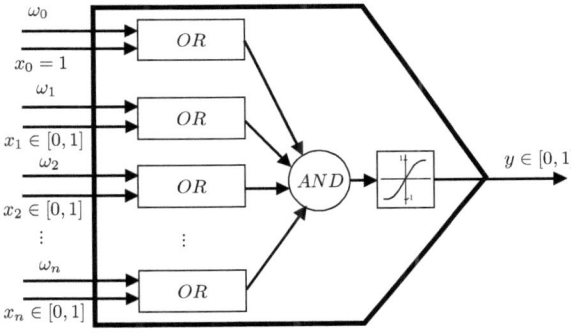

b: AND-OR-type fuzzy neuron

Figure 2.15: Types of the fuzzy neuron based on the operations

In the context of improvement of the fuzzy neural networks classifier, the concepts of the t-norm, t-conorm, and fuzzy implications were the development basis of various types of fuzzy neurons used in the (FNNs) such as t-conorm-based Mamdani-type neuro-fuzzy systems and fuzzy implications-based logical-type neuro-fuzzy systems [GGNZ06, Rut04].

A structure of the (FNNs) is similar to the structure of the (ANNs) (see Figure 2.4). Here the used neuron is replaced by a new type, the so-called *fuzzy neuron*, that is a neuron with the fuzzy uncertainties and the membership functions to express its input, weights, and also the activation functions used in the (FNNs) are similar to those used in the (ANNs) see (Figure 2.3) [Rut04, GGNZ06].

The types of the fuzzy neurons can be distinguished according to the logic operations as follows [Rut04, GGNZ06]

1. **OR-AND**-type fuzzy neuron (see Figure 2.15a) and is described as follows:
   $y = f(\text{OR}_{i=0}^{n} (\omega_i \text{ AND } x_i))$.

2. **AND-OR**-type fuzzy neuron (see Figure 2.15b) and is described as follows:
   $y = f(\text{AND}_{i=0}^{n} (\omega_i \text{ OR } x_i))$.

Maximum-margin fuzzy Classifier - Generally, the conventional fuzzy classifiers have a difficulty of adaptation due to the used design principle. The adaptation process of these classifiers means that the related fuzzy rules base should completely changed/redesigned in order to be consistent with the new changes related to the operation conditions of the considered system/process. This leads to the fact that the redesign of the related classifier will repeat from the beginning without any type of benefit of the information related to the classifier before changes of the operation conditions. Thus, the redesign process of the conventional fuzzy classifiers is strongly related to computational cost and time consuming problems [Abe10].

In order to avoid this problem, an idea of *trainable fuzzy classifier* is introduced. The trainable fuzzy classifiers can be distinguished according to shape of decision function regions into three types as

- fuzzy classifiers based on hyperbox regions,
- fuzzy classifiers based on polyhedral regions, and
- fuzzy classifiers based on ellipsoidal regions [Abe10].

The trainable fuzzy classifiers use one of the following procedures [Abe10]

- pre-clustering procedure
  Pre-clustering procedure can be achieved as

  - clusters are determined according to the own training data,
  - a fuzzy rule base is assigned by using the training data of the clusters, and
  - slopes and/or locations of the membership functions will be tuned in order to maximize the recognition rate of classifier using the given training data.

- post-clustering procedure
  Post-clustering procedure can be achieved as

  - for each class, the initial fuzzy rule is determined according to the own training data,
  - membership functions are tuned, and
  - based on the recognition rate of classifier using the given training data, new fuzzy rules can be defined.

For example, the design process of a conventional fuzzy classifier based on the ellipsoidal regions and the pre-clustering procedure is explained as follows [Abe10]

- center vector of the class i^{th} by

$$c_i = \frac{1}{M_i} \sum_{j=1}^{M_i} x_{ij}. \tag{2.17}$$

where (M_i) denotes the number of the training data for class i^{th} and (x_{ij}) is the jth training datum in the class i^{th}.

2.2 Fuzzy rule-based systems

- membership functions are defined by

$$\mu_i(x_j) = exp\left(h_i^2(x_{ij})\right), \qquad (2.18)$$

$$h_i^2(x_{ij}) = \frac{d_i^2(x_{ij})}{\alpha_i} = \frac{1}{\alpha_i}(x_{ij} - c_i)^T Q_i^{-1}(x_{ij} - c_i), \qquad (2.19)$$

where $(x_{ij} - c_i)$ is the distance between the datum (x_{ij}) and center vector of class i^{th}, Mahalanobis distance $d_i(x_{ij})$, $h_i(x_{ij})$ is tuned distance, α_i is the tuning parameter for class i^{th}, and Q_i is the covariance matrix for class i^{th} in the input space.

- the membership value of datum (x_{ij}) for each considered class is determined by using equation (2.18),

- the maximum membership value is calculated, as consequence the corresponding class of datum is defined (x_{ij}),

- the misclassification rate is evaluated to determine whether it needs to adjust all generated functions based on its related α_i parameter, and

- the adjustment process of α_i parameter related to each generated membership function will be repeated until getting the smallest percentage of misclassification, namely, the maximum generalization ability of designed classifier.

A design process of the Maximum-margin fuzzy classifier based on the ellipsoidal regions and the pre-clustering procedure is similar to the procedure of the design of the conventional fuzzy classifier based on the ellipsoidal regions and the pre-clustering procedure; but new steps should be added to adjust the parameter α_i as follows

- setting of the parameter α_i values for all membership functions of the classes to value equal to (1),

- adjustment of the parameter α_i to get the maximum margin of the membership function slope for all classes without causing the new misclassification,

- repetition of the adjustment of the parameter α_i within certain bounds so that the generalization ability is maximum as far as possible [Abe10].

Fuzzy min-max classifier A design process for a fuzzy min-max classifier is as follows
Due to distribution of classes/states associated with the training data, the *hyperboxes* are generated in the input space by a scanning process of the training data. Hereby the hyperboxes of the same class/state are allowed to overlap, whereas overlapping is not allowed between the hyperboxes of different classes/states. The principle of fuzzy min-max classifier is based on the *generation*, *expansion*, and *contraction* processes of the hyperboxes regions in the input space to get the maximum generalization ability as far as possible.

The training procedure, which can be characterized as *incremental training* process of the fuzzy min-max classifier can be summarized as follows

1. during the scanning process of the training data, the hyperbox is generated for each training sample. This hyperbox behaves in one of the two ways

    - in the first one, it expands within a predefined certain distance to unify with the other hyperbox for the same class.
    - in the second one, the generated hyperbox stays as it is because of the non-existence of any hyperbox including its state within the predefined certain distance.

2. overlapping between the hyperboxes of different classes is checked to achieve the contraction process to avoid the overlapping between them,

3. repetition of the previous steps to get a best discrimination between the hyperboxes of the different classes. Thus the generalization ability will be maximized [Abe01].

3 Adaptive Fuzzy-Based Approach (AFBA)

In this chapter, the adaptive fuzzy-based approach is explained in terms of the structure and basic concept.

3.1 Used terminology

In the context of this suggested approach, the following terms are used

- state (S)
  This term is understood as a linguistic expression or description distinguishing different conditions or operating states of the system to be considered. The related states are usually achieved by human classification or statements based on observations, study, and analysis of the considered system. So the expression "fault-free" represents the situation of regular operation of system and the status of operation in abnormal state can be expressed by "faulty". The total number of the related states of the considered system is denoted by M.

- feature (F)
  This term is understood as a synonym of attribute or an input variable to be able to highlight important relationships and underlying representations inside the raw signals with the related states in the best possible form of interpretation.

- hybrid
  This term denotes that an output of the feature extraction process consists of several types or different quantities of features.

- time series data
  The time series data are defined as sequence of data points, which are measured typically at successive time instants, spaced at uniform time intervals.

- structured data
  Structured data are data organized in a matrix structure whose rows contain the process features generated for the data and whose columns contain the data samples.

3.2 General structure

A block diagram of AFBA, which comprises training and modeling module and a classification module, is shown in Fig. 3.1 [AS10, AS11a, AS11b]. In the first module, the training data generated from the system are used to determine the M states considered. These states are determined based on human expertise, previous experience and/or operational observations. According to the data type, an additional feature extraction step may be applied. For time series or streamed data, a feature extraction process based on a sliding window is required to transfer the data into the feature space. The feature space introduces more detailed specifications with respect to the distinguishing property of the states considered; if the data are already structured, feature extraction is not necessary. These transferred time series or structured data are modeled using an embedded fuzzy-based process to generate a fuzzy model. The modeling process includes fuzzy partition, fuzzy partition selection, fuzzy rule generation and adaptation processes.

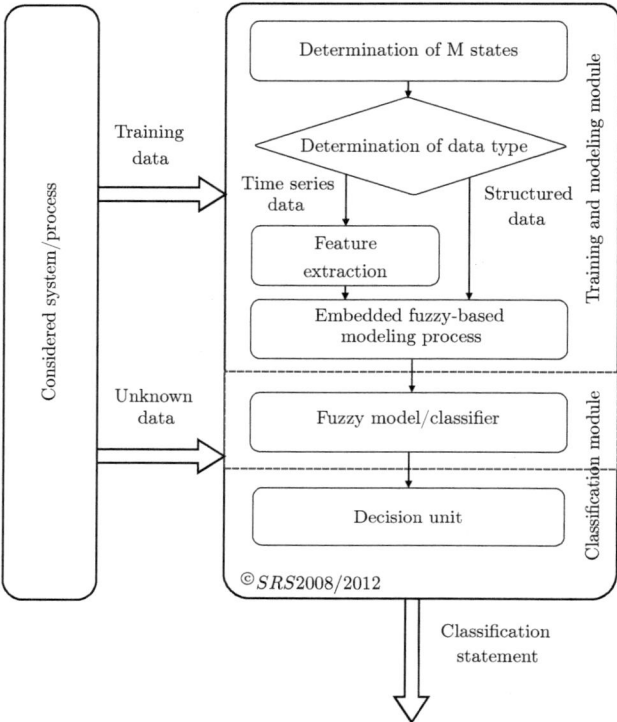

Figure 3.1: Basic structure of AFBA

Unknown or test data for the system are evaluated using the fuzzy model. The classification results are then integrated via the decision unit to generate a classification statement used to achieve further functionalities such as control, online evaluation and prediction tasks. The fuzzy model and decision unit are components of the classification module in our approach and the fuzzy model is a common component between the modules.

3.3 Feature extraction based on a sliding window

Measurement data in engineering applications are usually time series (offline) or streamed data (measured online without buffering and used for control or classification at the time of measurement). They can not be used directly as the input for pattern recognition approaches. Because of their specifications, these data can not be used to set decision boundaries to distinguish the states considered for the system.

The suggested approach generates suitable decision boundaries using feature extraction based on a sliding window concept (Fig.3.2).

The feature extractor transfers the data continuously into the feature space so that the states considered can be distinguished. The feature extractor can be adjusted using the first and the last desired parameter for the window, denoted SW_{BE} and SW_{EN}, respectively. Thus, a vector of time series or

3.4 Embedded fuzzy-based modeling

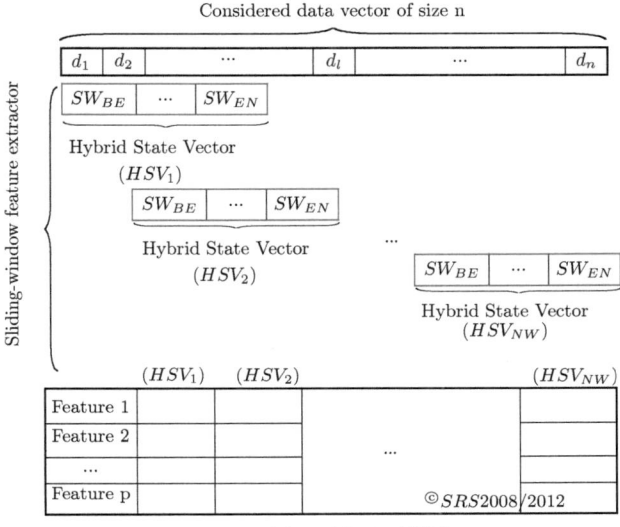

Figure 3.2: Feature extraction process using sliding windows

streamed data consisting of n samples can be scanned by NW subvectors, which continuously shift by one sample value. NW is calculated as

$$NW = (n - (SW_{EN} - SW_{BE})) + 1. \qquad (3.1)$$

Each subvector generates a new vector, a so-called hybrid state vector (HSV), consisting of p features. The statistical, mathematical, geometric and other features of the system can be used to build the HSV. The term "hybrid" indicates that the design of the state vector is based on different feature types and quantities. Finally, all the HSVs generated are combined to construct the hybrid state matrix (HSM) with p rows and NW columns [SAS08].

3.4 Embedded fuzzy-based modeling

Embedded fuzzy-based modeling (EFBM) is a key element of the proposed approach. EFBM includes the following steps

- Fuzzy partition/fuzzification process,
- Adaptation of the parameter boundaries,
- Fuzzy partition selection, and
- Fuzzy rule generation.

In combination with these steps, the concept of a homogeneity-oriented vector is introduced.

3.4.1 Homogeneity-oriented vector

In the design of pattern recognition approaches, the objective is to divide the feature space into several regions. These regions should be separable as far as possible to represent the states considered for the system. Separation is usually achieved by generating suitable decision boundaries. The performance of the design approach is strongly related to these decision boundaries. The boundaries should be carefully determined to avoid misclassification. The principle used for assignment has an important effect on the performance of the design approach. In the most widely used principle for assignment, the state of each range defined by generated decision boundaries corresponds to the state for the greatest number of data within the boundary. This principle does not hold for real applications because the samples are not easily separable in the real state or feature space for the system or process considered. In addition, application of this principle weakens the decision-making process because of misclassification and information loss problems.

To avoid these problems, a homogeneity-oriented vector (HOV) is proposed, defined as a weighting process for all states considered within the range of any range defined by generated decision boundaries, which can be derived from the fuzzy membership function here. HOV for the states in the fuzzy membership function within the feature range considered is defined as

$$HOV\left(\mu_F(S)\right) = \begin{bmatrix} \omega_{S_1} \\ \omega_{S_2} \\ \dots \\ \omega_{S_i} \end{bmatrix}, \qquad (3.2)$$

where $\omega_{S_i} = \frac{NoS_i}{TNoS}$ is a weighting factor for each state within the considered range, NoS_i is the number of samples classified as the ith state and $TNoS$ is the total number of samples, with $i = 1 : M$. The

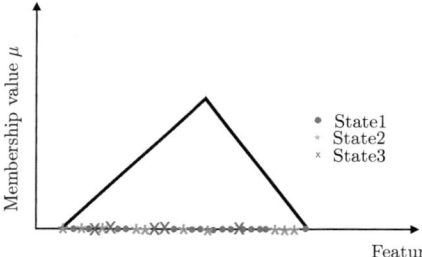

Figure 3.3: Example to explain the homogeneity-oriented vector

decision boundary representing the fuzzy membership function of Fig. 3.3 is used to explain the effects of HOV in the assignment process. According to the most widely used assignment principle, this function represents the first state because State1 has the largest number of samples within the function range. This principle leads to misclassification and information loss problems because any sample within the function range will be always classified into the State1, although the sample can actually belong to other states. Using the HOV concept, the function will represent all states to be included within the range as follows.

1. Calculation of the number of samples for each state, NoS_i, within the function range. In Fig. 3.3, the values of NoS_i for the three states considered are NoS_1=15, NoS_2=10, and NoS_3=5 samples.

2. Calculation of the total number of samples, $TNoS$, within the function range. In Fig. 3.3, the

3.4 Embedded fuzzy-based modeling

value of $TNoS$ is 30.

3. Calculation of the weighting factor w_{S_i} for each state. In Fig. 3.3, the values for the three states are $w_{S_1} = \frac{15}{30} = 0.5000$, $w_{S_2} = \frac{10}{30} = 0.3333$, and $w_{S_3} = \frac{5}{30} = 0.1667$.

4. HOV for the function in Fig. 3.3 is

$$HOV\left(\mu_F\left(S\right)\right) = \begin{bmatrix} 0.5000 \\ 0.3333 \\ 0.1667 \end{bmatrix} = \begin{bmatrix} w_{S1} \\ w_{S2} \\ w_{S3} \end{bmatrix}.$$

Therefore, HOV can guarantee that all information to be included within any decision boundary will be considered in the decision-making process. Thus, misclassification and information loss problems can be reduced as far as possible.

3.4.2 Fuzzy partition process

Fuzzy partition is an important stage in the design of fuzzy rule-based systems and affects the accuracy. The process includes determination of the fuzzy membership function to be used (such as triangular, trapezoidal or Gaussian).

The most widely used function is the triangular membership function.

The proposed approach uses a triangular fuzzy membership function (Fig. 3.4), defined as

$$\mu_F\left(x\right) = \begin{cases} 1 - (b-x)/(b-a), & \text{for } x \in [a,b] \\ 1 - (x-b)/(c-b), & \text{for } x \in [b,c] \\ 0, & \text{otherwise.} \end{cases} \qquad (3.3)$$

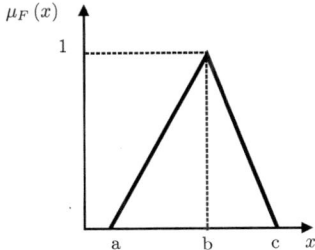

Figure 3.4: Triangular fuzzy membership function

In the terms of the suggested approach, the fuzzification process is achieved by using a developed technique, the so-called statistical characteristics-based technique.

The statistical characteristics-based technique used is as follows (Fig. 3.5 shows the procedure for two states, green for state 1 and red for state 2, within the range for feature F1):

1. A value vector of the samples is sorted in ascending order.

2. The mean $\bar{p}$ and standard deviation σ are calculated for the positions.

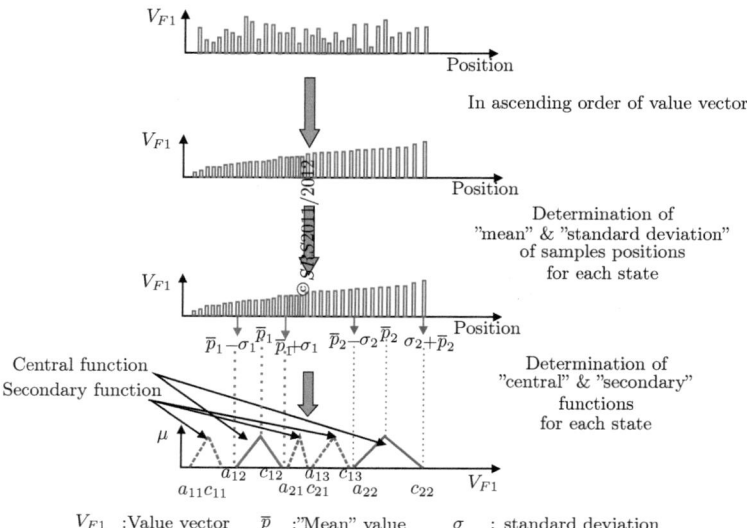

Figure 3.5: Statistical characteristics-based fuzzy partition technique within the range of feature (F1)

3. The first fuzzy partition, called central function, is limited by the $\bar{p} \pm \sigma$ positions for each state.

4. Other fuzzy partitions, called secondary functions, are limited by the remaining positions located outside the central function. The initial determination process of the secondary functions is arbitrarily realized.

5. The boundary parameters, namely a and c, for the central and secondary functions are defined according to the values for samples located in positions defined in steps 4 and 5.

The "position" term indicates the location place of each considered sample within the range of the training data vector.

Fig 3.6 illustrates the boundary parameters of the fuzzy partitions of two states within the range of three features (F1, F2, and F3). From this figure it can be seen that the number of the secondary functions can be different between the considered states within the range of individual feature (For example, within the range of the feature F1, the first state (green) needs the two secondary functions; while the secondary function of the second state (red) is one function). Also the number of the secondary functions can be different for the same state within the range of the different features (For example, the secondary functions of the first state (green) is two functions within the range of the feature F1 and the feature F2; while the number of these functions is one within the range of the feature F3). This difference is strongly related to the statistical and real distribution of the training data for each state and for each considered feature.

3.4.3 Adaptation of the boundary parameters in the feature space

The suggested fuzzy partition/fuzzification process (see Section 3.3.2) generates usually the functions fitting to the statistical and real distribution of the training data. However, these partitions can not be able to separate/distinguish very well the related states in the feature space. Therefore these partitions

3.4 Embedded fuzzy-based modeling

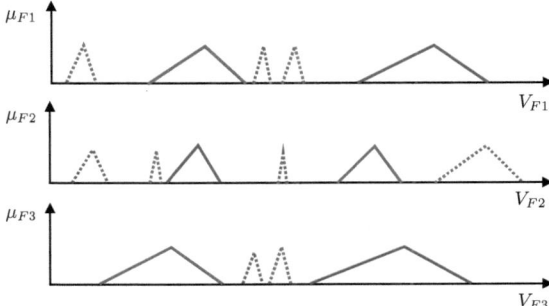

Figure 3.6: Initial values of the boundary parameters for the central and secondary functions

should be readjusted to achieve the fuzzy process with the best distinguish ability for the considered states.

The distinguish ability of each partition can be determined by means of the weighting factor values of HOV related to this function and the considered state. If the homogeneity of the considered fuzzy partition is bias for the considered state with the specific quality level, then this function is fitting to suitable separation of the considered states within the range of the feature space.

The quality level of the homogeneity related to any considered fuzzy partition and any considered state can be determined by the weighting factor value, namely ω_{S_i}, of the considered i^{th} state in the (HOV) related to this partition. The quality level of the homogeneity is often calculated as maximum possible for the considered state and the weighting factor values of the other states as minimum possible.

The adaptation/readjustment process plays an important role in the proposed approach and uses a two-dimensional adaptation criterion. This criterion is realized by adjusting the boundary parameters for the central and secondary functions for each state considered within the range for each suggested feature in two dimensions, as follows

1. Horizontal dimension

   A horizontal adaptation process aims to readjust the initial values of the boundary parameter(s) of the central function for each considered state within the range of each individual feature. The adaptation is realized to generate the new central and secondary functions, where the weighting factor value, namely ω_{S_i}, of the considered i^{th} state in the HOV related to these functions should become as large as possible and the weighting factor values of the other states as small as possible.

   The procedure for horizontal dimension adaptation is as follows

   - Initial values of the boundary parameters for the central and secondary functions are defined for each state within the range for each suggested feature using the fuzzy partition process in Section 3.4.2, as illustrated in Fig. 3.6.
   - Adaptation of the initial values of the boundary parameters for the central and secondary functions is realized as in Fig. 3.7.

     The determination process of which boundary parameter(s) to be shifted/readjusted is realized by using a checking process of what state of an adjacent fuzzy partition. If the state of the next function is the same then the related parameter is shifted/readjusted in the direction of this function (as in the case of the parameter a_{12} of the central function of the first state (green) (see last row in Fig. 3.5); while if the state of the next function is not from

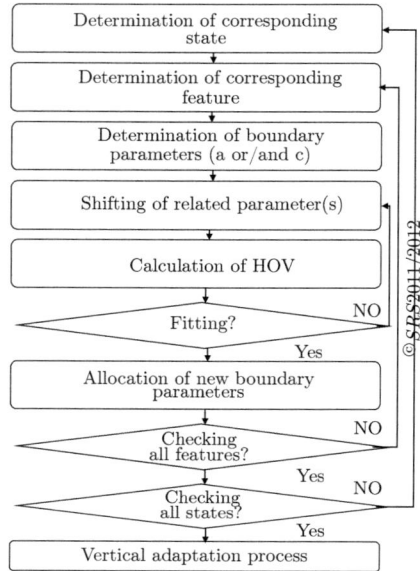

Figure 3.7: Adaptation of the initial values of the boundary parameters for the central and secondary functions (horizontal adaptation)

the same considered state, then the related parameter is not shifted/readjusted (as in the case of the parameter c_{12} of the central function of the first state (green)(see last row in Fig. 3.5). The calculation process of HOV is used as criteria to define whether this new fuzzy partition/function has better properties to distinguish the considered state or not. A determination process of the distinguish ability quality of each new function is denoted by "Fitting" process in Fig. 3.7.

The output of the horizontal dimension adaptation (Fig. 3.6) is illustrated in Fig. 3.8.

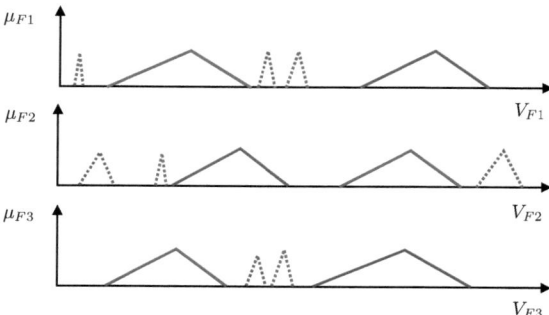

Figure 3.8: Output for horizontal dimension adaptation of the boundary parameters for the central and secondary functions generated in Fig. 3.6

3.4 Embedded fuzzy-based modeling

2. Vertical dimension

   The horizontal dimensional adaptation process ensures to readjust the boundary parameters within the range of each individual feature. But the suggested fuzzy classifier consists of the features combination. Therefore, it is necessary to achieve the additional readjustment process within this combination. This process is realized in the context of the suggested approach using a vertical dimensional adaptation process
   (Fig. 3.9).
   A checking concept of the overlapping region aims to reduce a complexity problem associated to the fuzzy rule base. The reduction process is based on a removing of the fuzzy partitions introducing same information with keeping the fuzzy partition, whose HOV is best between theses partitions, within the fuzzy partitions set related to each state. This concept ensures improvement of the accuracy with the reduction of the complexity problem. The output of the vertical dimensional

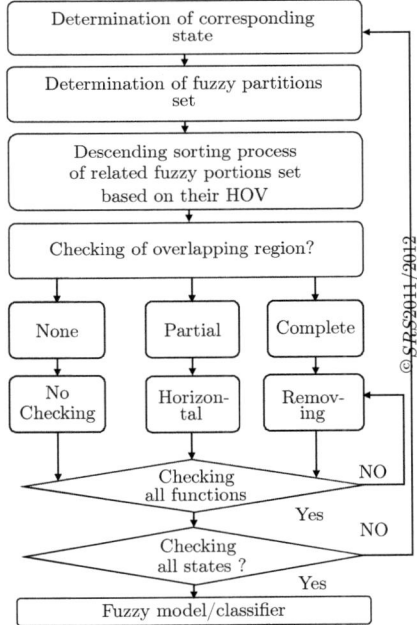

Figure 3.9: Vertical dimension adaptation

adaptation (Fig. 3.8) is illustrated in Fig. 3.10.

From Fig. 3.8 it can be seen that the feature F3 should be removed as the achievement of the two-dimensional adaptation process. Because the range of this feature can not include the decision boundaries to distinguish the related states.
Also it can be remarkable that there are the removed fuzzy partitions such as left secondary function of the first state (green) within the range of the feature F2, because this partition includes the information, which can be introduced by the positions of the central function of the considered state within the range of the feature F1. For avoidance of the repetition of the same information, the left secondary function of the first state (green) within the range of the feature F2 is removed. Also for the same reason, the

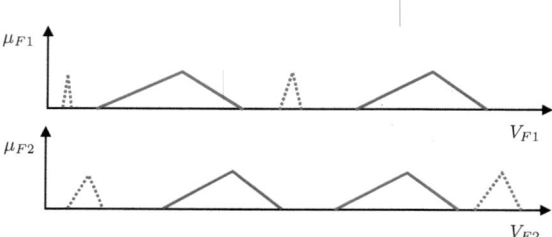

Figure 3.10: Output for vertical dimension adaptation of the boundary parameters for the central and secondary functions generated Fig. 3.8

central function of the considered state over the range of the feature F3 is removed.

The whole suggested two-dimensional adaptation process introduces an embedded principle in terms of the partition and feature selection process. Therefore, the approach is termed embedded fuzzy-based modeling.

3.4.4 Fuzzy rule generation and extraction

Fuzzy partition and adaptation generate a specific set of fuzzy partitions in the feature space to describe each state. These fuzzy descriptors should be combined in one format with all the relevant information. The number of fuzzy rules is equal to the number of states considered, denoted by M.

AFBA uses the format

$$R_j : \text{If } \cup_{k=1,l=1}^{K_j,L_j} x_{Fl} \text{ is } \mu_{F_{kl}}(S) \text{ with } HOV(\mu_{F_{kl}}(S)), \text{ then } y \text{ is } \begin{bmatrix} S_1 \\ \ldots \\ S_i \\ \ldots \\ S_M \end{bmatrix} \text{ with } \begin{bmatrix} CF_{j1} \\ \ldots \\ CF_{ji} \\ \ldots \\ CF_{jM} \end{bmatrix}$$

as the jth fuzzy rule to fuse all the relevant information for K_j fuzzy partitions within the range for L_j features.

The values of K_j and L_j differ between the states. The $\cup$ includes multiplication and summation processes used in the classification process (as illustrated in Fig. 3.11), $k = 1 : K_j$, $l = 1 : L_j$, $j = 1 : M$, $i = 1 : M$, x_{Fl} the value of lth feature of the input x, $\mu_{F_{kl}}(S)$ the kth fuzzy partition/antecedent/function within the range of lth feature, $HOV(\mu_{F_{kl}}(S))$ the kth homogeneity-oriented vector within the range of lth feature.

Here $\begin{bmatrix} CF_{j1} \\ \ldots \\ CF_{ji} \\ \ldots \\ CF_{jM} \end{bmatrix}$ is the vector of confidence factors for the states, where

$CF_{ji} = \frac{NoA}{K_j}$ is the confidence factor for the ith state based on all the antecedents for the jth rule and NoA is the number of antecedents to be included in the ith state within the range for the jth rule.

3.5 Fuzzy model

The suggested fuzzy model is based on a set of M fuzzy rules and evaluates the state of any sample using the following inference mechanism:

1. The p features of a sample are used as the input for the fuzzy model consisting of M fuzzy rules.

2. Each rule generates a number of fuzzy membership values equal to the number of fuzzy partitions or antecedents.

3.6 Classification process

The fuzzy model generates the number of fuzzy membership values for each of the input samples according to the number of rules and fuzzy partitions. The objective in the classification step is to obtain one fuzzy membership value for each state. These new values are used as a set to classify the final state of the input sample. To achieve this goal, the proposed approach uses a maximum operator based decision making process (Fig. 3.11). The procedure involves five steps:

1. Each fuzzy membership value is multiplied by the HOV of the antecedent/fuzzy membership function generating this value. The membership value itself can not be used to represent one state. This is because the corresponding antecedent or function usually includes many states within its range. Therefore, the multiplication process guarantees that this individual fuzzy membership value represents all possible states in range of corresponding antecedent/function, with percentages based on HOV values.

2. The values from step 1 are summed for each state to generate one classification value for each state within the range for each rule.

3. All classification values for the same state generated from all rules are summarized to generate one classification value for each state.

4. A maximum process is applied to the values from step 4 to define the final state of the input sample as a single output number.

The Σ_i process used in (Fig. 3.11) denotes the summation of all values representing the ith state generated in the previous step.

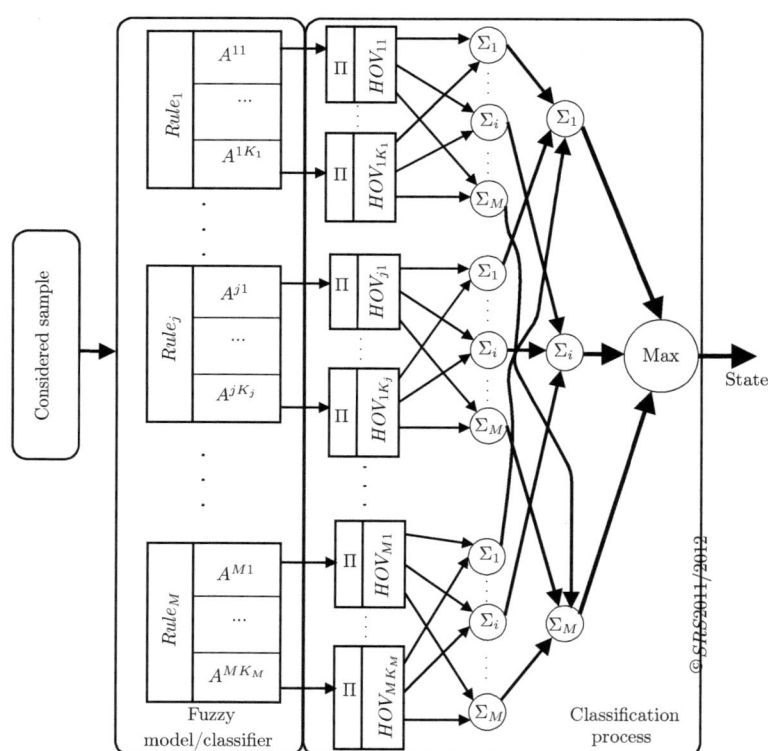

Figure 3.11: Fuzzy model and classification process.

4 Experimental validation of the new approach

After introducing of the AFBA approach in Chapter 3, this chapter shows the application and verification of the proposed approach as well as the comparison of the results with those of other approaches using benchmark data, using time series data, structured data and data resulting from a tribological system.

4.1 Implementation of the (AFBA) approach based on the benchmark datasets

4.1.1 Time series/streamed datasets

The proposed approach was evaluated using 12 benchmark sets from the UCR time series data set (Table 4.1.1) [KXWR06]. These data sets cover a wide variety of problems, ranging from two states to 7 states, for different applications areas, including biomedical, electromagnetic, and image processing.

The approaches used for comparison included k-nearest neighbor (k-nn), SVM, multi-layer perceptron (MLP; ANN method), Nets Bayes (NB; Bayesian classifier), C4.5 (decision tree classifier), logistic model tree (LMT), and Random Forests (RandForest), which are included in the Weka toolbox [KXWR06].

Name	#States	Length of signal	#Training data	#Test data
CBF	3	128	30	900
ECG200	2	96	100	100
Face (four)	4	350	24	88
Gun-Point	2	150	50	150
Lighting-2	2	637	60	61
Lighting-7	7	319	70	73
OSU Leaf	6	427	200	242
Synthetic control	6	150	50	150
Trace	4	275	100	100
Wafer	2	152	1000	6174
Coffee	2	286	28	28
Olive Oil	4	570	30	30

Table 4.1: Basic information for the benchmark data sets used

Percentage accuracy results for the benchmark data sets are shown for all the approaches, including the proposed AFBA, in Table 4.2. The percentage accuracy is defined as the ratio between the number of samples to be correctly classified by using the suggested approach and the total number of the considered samples. The results in Table 4.2 show clear differences in performance among the approaches for the benchmark data sets. The best approaches in terms of accuracy for each data set are shown in bold. To determine which approach is best in the terms of the whole used benchmark datasets, Friedman and Wilcoxon signed rank tests will be applied. The Friedman ranking process aims to compare the performance of the used approaches within the range of the complete dataset; while the Wilcoxon signed rank test is used to compare their related performance within the range of each individual dataset.

Data set	Approach							
	k-nn	NB	C4_5	MLP	Rand-Forest	LMT	SVM	AFBA
CBF	85.00	**89.67**	67.33	85.33	83.56	77.00	87.67	89.89
ECG200	89.00	77.00	72.00	84.00	81.00	82.00	81.00	**94.00**
Face (four)	87.50	84.09	71.59	87.50	78.41	77.27	88.64	**89.77**
Gun-Point	92.00	78.67	77.33	**93.33**	89.33	79.33	80.00	91.33
Lighting-2	80.33	67.21	62.30	73.77	78.69	63.93	72.13	**81.97**
Lighting-7	63.01	64.38	54.79	64.38	56.16	64.38	**71.23**	67.12
OSU Leaf	**54.55**	37.19	36.78	44.63	41.74	49.17	43.80	52.89
Synthetic control	88.00	**96.00**	81.00	91.33	86.00	92.00	92.33	94.67
Trace	82.00	80.00	74.00	77.00	81.00	76.00	73.00	**99.00**
Wafer	**99.40**	70.83	98.20	96.28	99.32	98.09	95.96	99.11
Coffee	75.00	67.86	57.14	96.43	75.00	**100**	96.43	**100**
Olive Oil	76.67	76.67	73.33	86.67	86.67	83.33	86.67	**90.00**
Mean	81.03	74.13	68.81	81.72	78.07	78.54	80.73	**87.47**

Table 4.2: Accuracy values of the AFBA approach and the comparative approaches applied to the used benchmark datasets

The Friedman test involves checking a null hypothesis that states that all approaches are equivalent and so their ranks should be equal. A p value is calculated and the null hypothesis is true if p is greater than 5% (significance level); otherwise, the hypothesis is rejected. When the Friedman test is applied to the results in Table 4.2, the results $p = 0.00002$ is ≤ 0.05, so the null hypothesis is rejected. Friedman's ranking of the approaches according to the results in Table 4.2 is shown in Table 4.3.

The Wilcoxon signed rank test is similar to the Friedman test in terms of checking a null hypothesis. The results are shown in Table 4.5 and Table 4.4. According to the Wilcoxon signed sum test, the $p = 0.0078$ value is less than the significance level of 0.05. Thus, the null hypothesis is rejected.

An increasing of the ranking value in Table 4.3 and Table 4.4 denotes that the related approach is best. According to results of Table 4.3 and Table 4.4, the suggested approach is best approach in the terms of the whole evaluation results of the used benchmark datasets.

Approach	Ranking
AFBA	13.2083
MLP	9.9167
k-nn	9.6250
SVM	9.1667
RandForest	8.0000
LMT	7.8333
NB	6.5833
C4_5	3.6667

Table 4.3: Average ranking for the comparative approaches for all of the data sets according to Friedman test (highest is best)

4.1.2 Structured datasets

In this section the application results of the proposed approach to structured datasets are shown. In the context of this contribution, structured data are data organized in a matrix structure whose rows contain the process features generated for the data and whose columns contain the data samples. The basic information of the used benchmark structure datasets are given in Table 4.6 [AFFL+11]. Each

4.1 Implementation of the (AFBA) approach based on the benchmark datasets

Approach	Ranking
AFBA	7.3750
k-nn	5.4583
MLP	5.1667
SVM	4.7500
RandForest	4.1667
LMT	4.0417
NB	3.6250
C4_5	1.4167

Table 4.4: Average ranking for the comparative approaches for all the benchmark data sets according to Wilcoxon test (highest is best)

Data set	Approach							
	k-nn	NB	C4_5	MLP	Rand-Forest	LMT	SVM	AFBA
CBF	4	7	1	5	3	2	6	**8**
ECG200	7	2	1	6	3.5	5	3.5	**8**
Face (four)	5.5	4	1	5.5	3	2	7	**8**
Gun-Point	7	2	1	**8**	5	3	4	6
Lighting-2	7	3	1	5	6	2	4	**8**
Lighting-7	3	5	1	5	2	5	**8**	7
OSU Leaf	**8**	2	1	5	3	6	4	7
Synthetic control	3	**8**	1	4	2	5	6	7
Trace	7	5	2	4	6	3	1	**8**
Wafer	**8**	1	5	3	7	4	2	6
Coffee	3.5	2	1	5.5	3.5	**7.5**	5.5	**7.5**
Olive Oil	2.5	2.5	1	6	6	4	6	**8**

Table 4.5: Average ranking for the comparative approaches for each individual data set according to Wilcoxon test(highest is best)

Name	#Classes	#Features	#Samples	IR
Iris	3	4	150	1 (Balanced)
PID	2	8	768	1.87 (Imbalanced)
Sonar	2	60	208	1.15 (Imbalanced)

Table 4.6: Basic information of the benchmark structure datasets

benchmark data set is described as follows.

1. Iris dataset
   This data set contains 50 samples for each state from three classes (Iris setosa, Iris versicolor, and Iris virginica). The related features are four continuous features as follows
   - sepal length,
   - sepal width,
   - petal length, and
   - petal width.

2. PID dataset
   This data set contains 500 samples for the first class (diabetic) and 268 samples for the second class (healthy). The related features are eight features (as shown in Table 4.7).

3. Sonar dataset

    This data set is characterized by a high-dimensional property. The data set contains 111 and 97 samples from two classes, i.e., sonar signals from mine (metal cylinders) (class 1) or rocks (class 2), respectively. The related features are 60 features.

The datasets are chosen for the evaluation of the AFBA approach due to

- Variety in high-dimensional problems such as large (Sonar dataset), medium (PID datasets) and small (Iris dataset) and

- Variety in values with respect to their *Imbalance Ratio*. The Imbalance Ratio (IR) is defined by the ratio between instances/samples of the majority state and the minority state.

Feature number	Feature name
1	Number of thimes pregnant
2	Plasma glucose concentration
3	Diastolic blood pressure (mm Hg)
4	Triceps skin fold thickness (mm)
5	2-Hour serum insulin (mu U/ml)
6	Bodu mass index
7	Diabetes pedigree function
8	Age

Table 4.7: Features for the PID dataset

To develop different experiments, the tenfold-cross validation model is used. The cross validation model divides the considered benchmark dataset into ten folds, whereby nine folds are used as training data and the remaining fold is used for the testing phase. In the context of the evaluation phase of the AFBA approach, the tenfold-cross validation process is iterated three, five, and ten times, namely, 3-10cv, 5-10cv, and 10-10cv, respectively. As final result, the average value for the generative results during the corresponding iterative process is given.

The main reason to use this model and this iteration process is to avoid a bias problem. The proposed method will be tested randomly and according to the different conditions.

In the sequel, the evaluation results of the structured benchmark datasets based on the AFBA approach as well as the comparison of the results to those of other approaches are introduced.

1. "Iris" dataset

    The evaluation and comparison results using the "Iris" dataset using 3-10cv, 5-10cv, and 10-10cv, respectively, are given in Table 4.8, Table 4.9, and Table 4.10.

2. "Pima Indians Diabetes (PID/Pima)" dataset

    The evaluation and comparison results using the "PID/Pima" dataset using 3-10cv, 5-10cv, and 10-10cv, respectively, are given in Table 4.11, Table 4.12, and Table 4.13.

3. "Sonar" dataset

    The evaluation and comparison results using the "Sonar" dataset using 3-10cv, 5-10cv, and 10-10cv, respectively, are given in Table 4.14, Table 4.15, and Table 4.16.

The comparison process of these results focuses to the relationship between the accuracy level and the number of rules associated with each proposed approach. This dependency determines, which priority quality level for each part (accuracy and number of rules) is realized during the design process.

The priority quality level can be one of the following cases.

4.1 Implementation of the (AFBA) approach based on the benchmark datasets

	FARC-HD ([AFAH11])	Product 1-2-3 ([ANHI09])	AFBA
Accuracy	**96.00**	95.33	**96.00**
#Rules	4	5.23	**3**

Table 4.8: Evaluation results for AFBA compared to other approaches using iris dataset and 3-10cv

	H_{100} ([IKN08b])	SGERD ([MZK08])	AFBA
Accuracy	96.13	**96.93**	96.53
#Rules	300.0	3.96	**3**

Table 4.9: Evaluation results for AFBA compared to other approaches using iris dataset and 5-10cv

	FMM-GA ([QLT10])	AFBA
Accuracy	**98.27**	95.13
#Rules	7	**3**

Table 4.10: Evaluation results for AFBA compared to other approaches using iris dataset and 10-10cv

	FARC-HD ([AFAH11])	All granularities ([ANHI09])	AFBA
Accuracy	**75.66**	74.92	59.52
#Rules	22.7	6.63	**2**

Table 4.11: Evaluation results for AFBA compared to other approaches using PID/Pima dataset and 3-10cv

	SGERD ([MZK08])	AFBA
Accuracy	**74.64**	57.53
#Rules	6.12	**2**

Table 4.12: Evaluation results for AFBA compared to other approaches using PID/Pima dataset and 5-10cv

	FMM-GA ([QLT10])	[AZ08]	AFBA
Accuracy	**89.74**	74.71	56.36
#Rules	37	5.36	**2**

Table 4.13: Evaluation results for AFBA compared to other approaches using PID/Pima dataset and 10-10cv

	FARC-HD ([AFAH11])	AFBA
Accuracy	**80.19**	63.49
#Rules	18	**2**

Table 4.14: Evaluation results for AFBA compared to other approaches using sonar dataset and 3-10cv

1. priority for accuracy

    In this case, the accuracy part is prioritized regardless the number of rules part. In the context of this case, the proposed approach has two properties to be detailed.

    - balanced datasets

	SGERD ([MZK08])	AFBA
Accuracy	74.80	62.00
#Rules	4.92	2

Table 4.15: Evaluation results for AFBA compared to other approaches using sonar dataset and 5-10cv

	FMM-GA ([QLT10])	[AZ08]	AFBA
Accuracy	98.45	74.90	59.90
#Rules	12	7	2

Table 4.16: Evaluation results for AFBA compared to other approaches using sonar dataset and 10-10cv

The proposed approach shows promising results (see Tables [4.8-4.10]).

- imbalanced datasets

  The related results of the proposed approach are not satisfying in comparison with other approaches (see Tables [4.11-4.16]).

It can be concluded that the proposed approach has a strong sensitivity for the imbalance ratio value of the considered dataset.

2. priority for number of rules

   In this case, the rules number part is desired regardless the accuracy part. Here, the proposed approach shows better results than other approaches because the number of the extracted rules is always equal to the related states; also this number is independent of the parameters of the used classification technique (see Tables [4.8-4.16]).

3. priority for both

   In this case, the accuracy and the number of rules parts are desired. In the context of this case, the proposed approach shows acceptable results. The generated accuracy level with the used number of rules is acceptable in comparison with other approaches.

   For example, from Table 4.10, it can be noticed that the first approach [QLT10] needs seven rules to realize 98.27% as accuracy value; while the proposed approach needs three rules to get 95.13% as accuracy value. The first approach [AFAH11] in Table 4.11 needs 22 rules to get 75.66% as accuracy value; while the proposed approach needs two rules to get 59.52% as accuracy value, and so on.

Consequently, as future aspect the proposed approach could be modified to reach the satisfied degree for the interactive relationship between the accuracy and the rules number parts.

4.2 Application of AFBA approach to tribological system

In the following section, the suggested approach is evaluated and verified using time series/streamed dataset generating from mechanical system. The considered mechanical system is understood as the tribological system. The tribology term is understood as a theoretical framework defining the interactive relationships between surfaces due to relative movements. The study and application of principles of *friction*, *lubrication*, and *wear* are covered using the tribology science and engineering [Hut92].

4.2 Application of AFBA approach to tribological system

4.2.1 Test rig

A test rig illustrated in Figure 4.1 is designed to mimic the real tribological system [Det11]. In order to realize operation conditions related to considered field application, the dimensions of components used in this test rig are carefully determined.

The considered test rig consists of the following parts

1. two wear plates,

2. body locked into position, and

3. counter body connected to the cylinder.

A linear movement of washboard profile of wear plates limits a horizontal movement. While a vertical movement is prevented by an adjustable contact pressure, realized by a pneumatic cylinder and a lever arm. A lubrication process is automatically realized at fixed time intervals. Typical measurements from field application (normal force, acceleration, hydraulic pressure, temperature, etc.) are recorded by the real-time system, which is not depicted here. As operation phases, the counter body strokes within a load phase with length 40s and subsequently pauses for 70s. This cycle of 110s is repeated steadily.

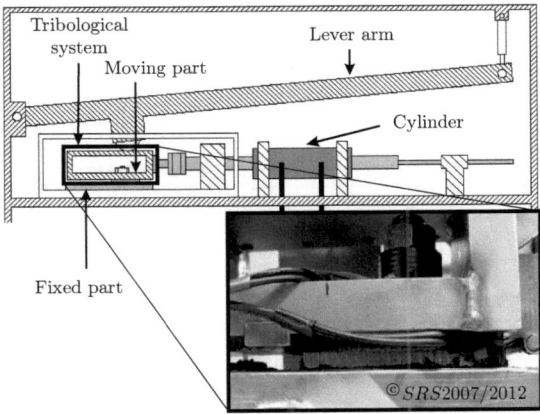

Figure 4.1: Test rig as the tribological system

4.2.2 Problem statement and hypothesis

A test rig is used for studying/analysis the erosion rate-based friction and wear processes of metal surface (see Figure (4.2)) related to changes of conditions of operation such as changes of lubrication and temperature, etc. based on analysis of the pressure, force, and acceleration signals generating from this system [Det11]. According to the results, the best signal being is determined and used to build the classifying model. Therefore, temperature, pressures, and force as well as acceleration signals generated from the system can be used to be analyzed. A suggested hypothesis of behavior of the metal surface based on the erosion rate illustrates in Figure 4.3. This behavior consists of three regions as follows

Figure 4.2: Example of erosion rate of the metal surface

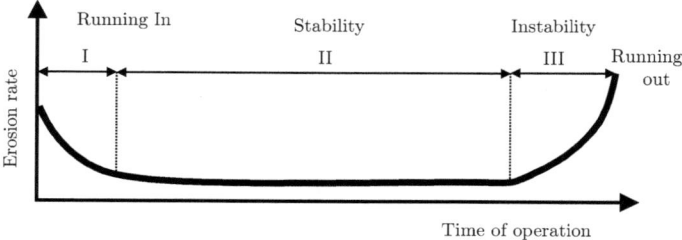

Figure 4.3: Hypothetical behavior of the erosion rate during the operation time

1. running in region (I)

   This region represents a beginning of the operation. The changes related to this region can contain all related states of the considered operation.

2. stable region (II)

   This region includes the acceptable level of the erosion rate of the metal surface. If there is any change in the erosion rate, this change does not take long time to come back to the stable states.

3. unstable region (III)

   The region indicates that the erosion rate of the metal surface should be carefully observed in order to make correct decision before reaching to the running out moment.

4.2.3 Training and modeling phase

All related examinations between surface conditions, changing of operation conditions, and all typical measurements from field application have shown that the pressure signal is a suitable signal to be used for further examinations about the changes of the surface conditions. Additionally, based on the evaluation of all related experiments, the five/four related states ($M = 5$) of surface conditions should be distinguished

(Table 4.17) [AS11b]. The 34 statistical feature set are used to build the hybrid state matrix (HSM) for each operation cycle. The 12 features are selected from this set to build the desired fuzzy model/classifier.

Nr.	Linguistic expression	Color representation	Human classification
1	Regular 1	Green	Stable operation
2	Regular 2	Blue	Stable operation with minor changes
3	Regular 3	Cyan	Stable operation with acceptable changes
4	Regular 4	Yellow	Abrupt changing surface conditions with acceptable changes
5	Regular 5	Red	Abrupt changing surface conditions with non-acceptable changes

Table 4.17: Overview of the related states of surface conditions/erosion rate the oil lubrication

4.2.4 Representations spaces

For representation of the proposed approach-based evaluation/analysis results of each operation cycle, the following spaces are suggested

1. fuzzy classifier output space

    A building of fuzzy classifier output space is based on a vector, so-called *Fuzzy Classifier Output Vector*, in short FCOV, for each operation cycle. This vector consists of M values related to all considered states. The FCOV indicates the classification of the corresponding cycle based on the suggested fuzzy classifier/model of M related states.

    The objective of this space is to highlight generally which state is dominant for each operation cycle. The dominant state is corresponding to the state, which has the maximum numerical value of the FCOV. However, this representation is only useful to make decision for classification and diagnosis goals; but in the terms of the tasks and goals of this tribological system, it necessarily needs more details information about the state of the erosion rate to make correct decision (such as setting to the suitable condition operation or change of the sheet). The fuzzy classifier output space is not the best representation for achieving this goal.

    The operation cycle number 211 in Figure 4.4 is used to clarify the above-mentioned idea. In the fuzzy classifier output space, the FCOV of this cycle is

$$\begin{bmatrix} FCO_{S5} \\ FCO_{S4} \\ FCO_{S3} \\ FCO_{S2} \\ FCO_{S1} \end{bmatrix} = \begin{bmatrix} 0.0042 \\ 0.2543 \\ 0.3371 \\ 0.2905 \\ 0.3645 \end{bmatrix}. \tag{4.1}$$

According to FCOV, the state of the corresponding cycle is within the range of the stable operation, namely $S1$. But here in this application, the goal is to define a percentage of the erosion rate of the metal surface. From the FCOV of this corresponding cycle, it can be noticed that there is a type of convergence between the first four states, which indicates basically different percentage of changes of the erosion rate of the metal surface. Thus, it is not advisable to say that the state of the corresponding cycle is $S1$. For avoidance of this conflict, the concept of the state and spectrum spaces are introduced.

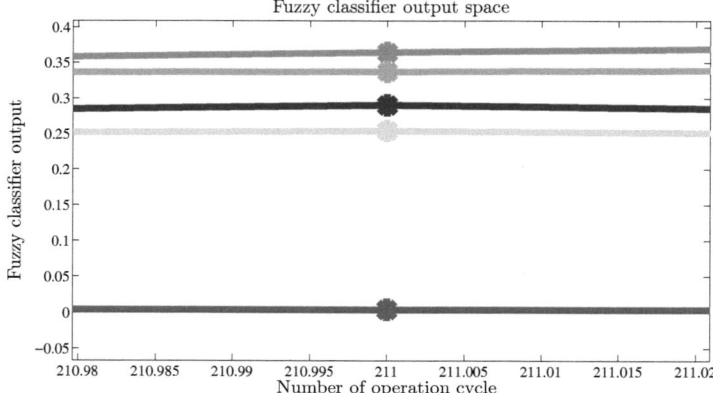

Figure 4.4: Representations of the operation cycle number 2011 based on fuzzy classifier output space

2. state space

As it is earlier mentioned, the fuzzy classifier output space is strongly related to the conflict problem for determination the percentage of the erosion rate of the metal surface. The representation of state space is therefore suggested for getting rid this problem.

The state space includes the M lines represented the related states of the considered system (here, i.e. state space with 5 lines (see Figure 4.5)). A building of this space is based on the following hypotheses

(a) *"the erosion rate increases due to the increase of damages of the metal surface combined with the continued operation"*

(b) *"a sequence of the color representation/coding of the related states is from state 1 to state M"*

The representation procedure of each operation cycle within the state space is as follows

(a) determination of the state number corresponding to maximum value of FCOV.
In the case of the example of the cycle number 2011, the corresponding state is $S1$ because this state has maximum value of FCOV, namely 0.3645, (see 4.1).

(b) determination of number of states to be lower than the state to be defined in the first step.
In the case of the example of the cycle number 2011, there is no state lower than $S1$ (see 4.1), because this state represents the first state in this space according the second hypothesis.

(c) calculation of summation of the FCO values of the states to be defined in the second step.
In the case of the example of the cycle number 2011, the summation value is equal to 0.000 (see 4.1).

(d) determination of number of states to be higher than the state to be defined in the first step.
In the case of the example of the cycle number 2011, the number of the states to be higher than $S1$ is four states $[S2 - S5]$ (see 4.1).

4.2 Application of AFBA approach to tribological system

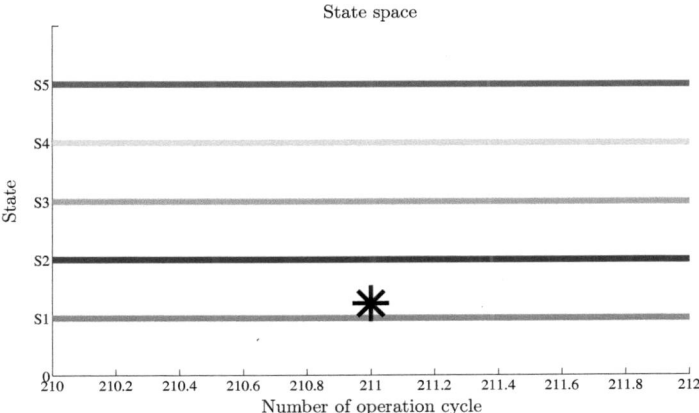

Figure 4.5: Representations of the operation cycle number 2011 based on State Space

(e) calculation of summation of the the FCO values of the states to be defined in the fourth step.
   In the case of the example of the cycle number 2011, the summation value is equal to $FCO_{S2} + FCO_{S3} + FCO_{S4} + FCO_{S5} = 0.2905 + 0.3371 + 0.2543 + 0.0042 = 0.8861$ (see 4.1).

(f) subtraction process of the values generated from the first step and the third step.
   In the case of the example of the cycle number 2011, $0.3645 - 0.0000 = 0.3645$.
   The $(-)$ sign denotes the FCO values of these state will pull the value of the state defined in the first step toward a new position to be lower than the old position.

(g) summation process of the values generated from the fifth step and the sixth step.
   In the case of the example of the cycle number 2011, $0.3645 + 0.8861 = 1.2506$.
   The $(+)$ sign denotes the FCO values of these state will pull the value of the state defined in the sixth step toward a new position to be higher than the old position.

(h) painting the value generated from the previous step in the state space.

This suggested procedure is applied to the example of the operation cycle number 211 in Figure 4.4 to generate a value of 1.2506.

The representation of the state space introduces a visualizing framework of all available information generated from several sources, here the several sources are the fuzzy model/classifier consisting M fuzzy rules, about the same operation cycle.

3. spectrum space

   A goal of spectrum space is similar to the framework of the state space in the visualizing of all available information generated from several sources. But in this space, the final value is represented by the color visualization instead of the numerical visualization as in the state space.

   The procedure of the color visualization in the spectrum space is as follows

   (a) determination of the final value from the representation procedure of each operation cycle within the state space.

   (b) determination of the colors of two boundaries of subspace, wherein the final value is located in the state space.

(c) determination of a color scale related to the colors to be defined in the second step.

(d) determination of a color of the considered final value based on the color scale to be defined in the third step and the value to be located after a decimal point in this value.

(e) painting of the final value as rectangle in the spectrum space with the color corresponding to the color defining in the fourth step.

The color visualization of the operation cycle number 211 in Figure 4.6 is the rectangle with the color scale $[red\ green\ blue] = [0\ 0.7496\ 0.2506]$ because this the numerical representation of this cycle in the state space located between $S1$ line (represented by green color) and the $S2$ line (represented by blue color).

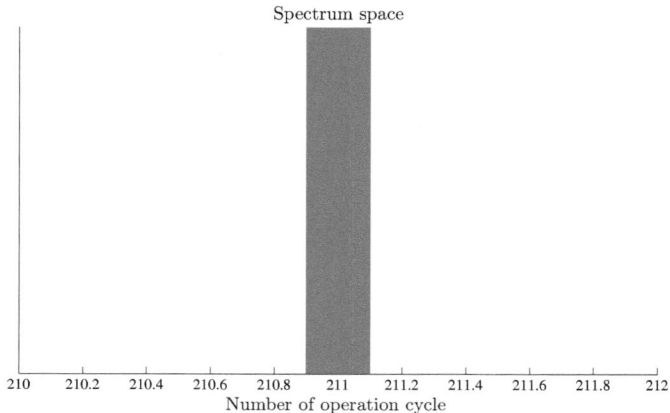

Figure 4.6: Representations of the operation cycle number 2011 based on Spectrum Space

4.2.5 Testing and evaluation phase

To test the suggested algorithm, two datasets are used. According to the human classification, an erosion rate of these datasets is as follows

1. in the first dataset, the erosion rate increases combined with the continued operation.
2. the erosion rate of the second dataset is within regular operation range.

The evaluation results of the first dataset are illustrated in Figure (4.7), Figure (4.8), and Figure (4.9).

As it can be seen from Figure (4.7), a gradual disappearance of the related colors, namely green, blue, and cyan colors, presenting the states of regular 1, regular 2, and regular 3 of erosion rate and a gradual appearance of colors, namely yellow and red colors, representing the states of abnormal 1 and abnormal 2 of erosion rate changes with the progress of run-time.

Additionally, an approximate and continuously transition of the states from states 1, 2, and 3 to states 4 and 5 can be seen from Figure (4.8).

The spectrum space of the evaluation results for the first dataset demonstrates that the color change related to the change of the erosion rate transits gradually form the area of the stable operation into the unstable area with the progress of run-time (see Figure 4.9).

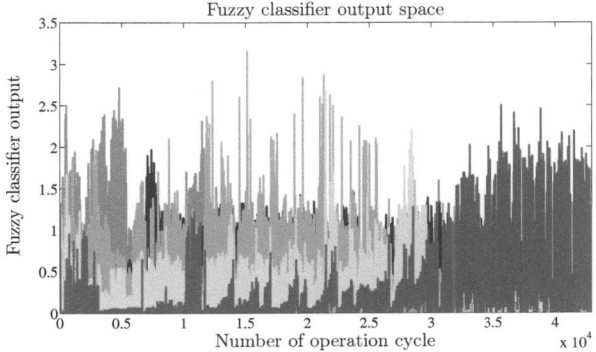

Figure 4.7: Evaluation results of the first dataset the fuzzy classifier output space

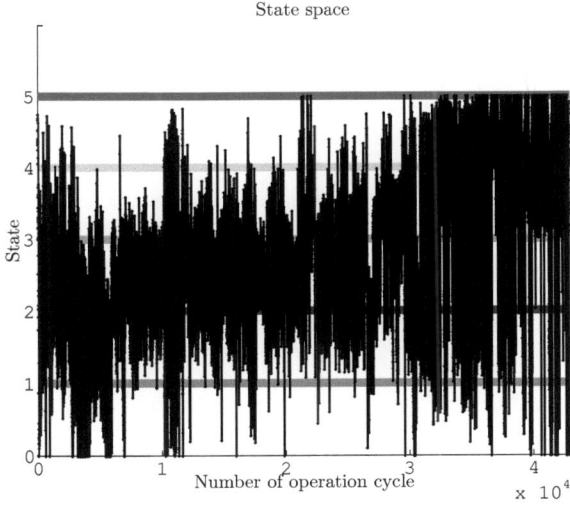

Figure 4.8: Evaluation results of the first dataset in the state space

The evaluation results of the second dataset in Figure (4.10), Figure (4.11), and Figure (4.12). These results are also consistent with the human classification to the changes of states for this data, which can also be seen by the complete appearance of the colors green, blue and cyan colors, represented the states of regular 1, regular 2, and regular 3 of erosion rate in Figure (4.10). From Figure (4.11) and Figure (4.12), it is observed that the change of state is approximately between the states 1, 2, and 3.

The evaluation results show that there approximately is consistence between the theoretical hypothesis (see Figure 4.3) and the applicable hypothesis (see Figure [4.7- 4.12]) about the behavior of the erosion rate with the progress of the operation.

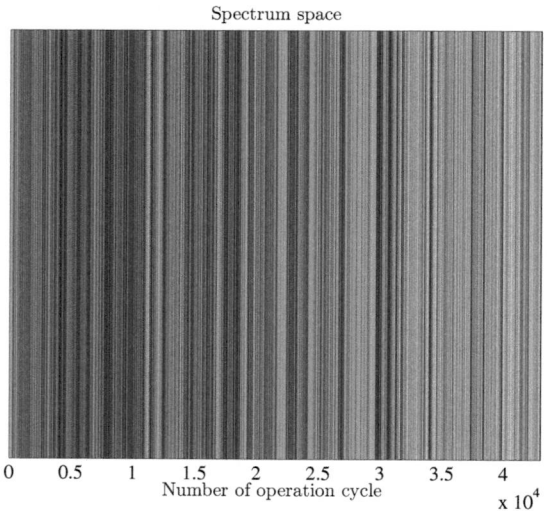

Figure 4.9: Evaluation results of the first dataset in the spectrum space

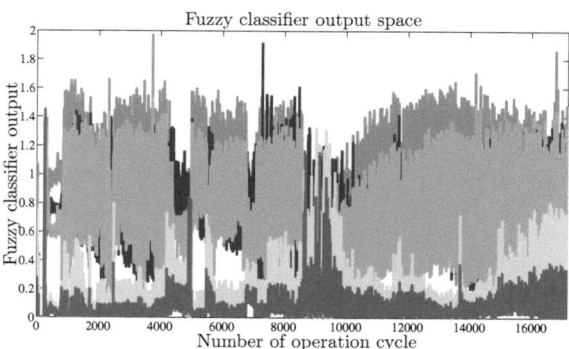

Figure 4.10: Evaluation results of the second dataset the fuzzy classifier output space

4.2 Application of AFBA approach to tribological system

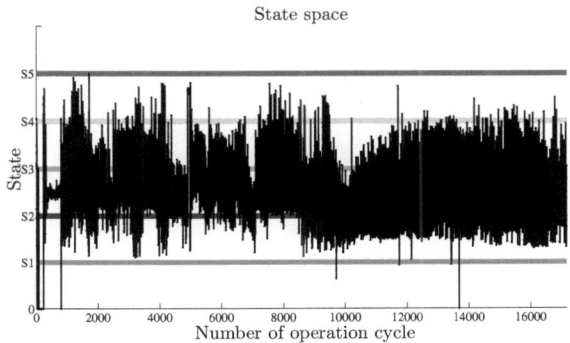

Figure 4.11: Evaluation results of the second dataset in the state space

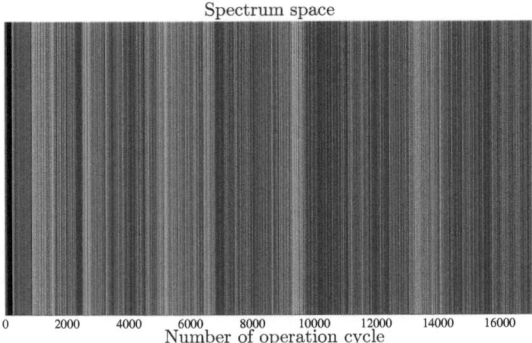

Figure 4.12: Evaluation results of the second dataset in the spectrum space

5 Summary and conclusions

In the terms of the wide usage of systems based on fuzzy rules the development of pattern recognition, machine intelligence, and control systems, this work focuses on the development of a fuzzy rule-based system.

In the present work, the suggested adaptive fuzzy-based approach (AFBA) is based on the following new aspects

1. automated and improved generation of fuzzy rules based on the statistical properties of the data considered and

2. automated generation of features to be more informative for the state related to the data considered.

This work consisted of the following chapters. The chapter 1 introduced the state of art of non-fuzzy-based pattern recognition approaches, pattern recognition approaches of time series data, and fuzzy-based pattern recognition approaches in the context of the current techniques and the related advantages and disadvantages. The chapter 2 presented the pattern recognition and fuzzy rule-based systems in the terms of the basic concepts, structure, and application. The proposed approach is introduced in the chapter 3 in the context of the following sections

1. general structure,

2. automatic and adaptive sliding-window feature extraction process,

3. embedded fuzzy-based modeling process including,

    - homogeneity-oriented vector-based assignment process for overcoming the loss information and misclassification problems,
    - statistical characteristics-based fuzzy partition process,
    - two-dimensional adaptation process for improvement of the accuracy and performance of the fuzzy rule-based systems, and
    - fuzzy rule generation and extraction.

4. fuzzy model and classification process.

The proposed approach was evaluated using the benchmark time series and structure datasets and was compared by the comparative approaches. As real application, the approach suggested was applied to the tribological system. All results of the evaluation, comparison, and application were shown in the chapter 4. According to these evaluation results, the suggested approach performance can be summarized as follows

1. the benchmark time series dataset

    The suggested approach is competitive and comparable to the standard approaches such as ANNs, SVM, k-nn, etc (see Table 4.3 and Table 4.4). Additionally the suggested structure of AFBA introduces the suitable framework for building of the pattern recognition technique without need to predefine the modeling parameter such as (number of layers, of neurons, the initial values of the weights vector in ANNs or margin control parameter and the number of support vectors in SVM or the number of clusters in k-nn). Whereby, in the context of this approach, the determination and adjustment processes of the modeling parameters (Here number of the fuzzy triangular membership functions and the related their control parameters (a, b, and c parameters)) were realized based on the statistical properties of the data considered.

2. the benchmark structured dataset

   The performance of AFBA can be distinguished to priority type in the interactive relationship between the accuracy level and the rules number.

   (a) priority for the accuracy

   In this case, the accuracy part is desired regardless of the rules number part. In the context of this case, the proposed approach has two behaviors as follows

   - balanced datasets
     The proposed approach shows the promising results (see Tables [4.8-4.10]).
   - imbalanced datasets
     The related results of the proposed approach are not satisfied in comparison with the other approaches (see Tables [4.11-4.16]).

   (b) priority for the rules number

   In this case, the rules number part is desired regardless of the accuracy part. In the context of this case, the proposed approach shows the results better than the other approaches because the number of the extracted rules is always equal to the related states and also this number is independent of the parameters of the used classification technique (see Tables [4.8-4.16]).

   (c) priority for both

   In this case, the accuracy and rules number parts are desired. In the context of this case, the proposed approach shows the results to be acceptable because the generated accuracy level with the used rules number is acceptable in comparison with the other approaches.

3. tribological system

   The AFBA-based automated evaluation results of the considered datasets showed the ability of this approach to classify/determine the state of the erosion rate of the metal surface at each operation cycle. These results showed the consistence type between the human evaluation and the automated evaluation for the erosion rate of the metal surface. Also this approach introduced the visualizing framework, namely state and spectrum spaces, of all available information generated from several sources, here the several sources are the fuzzy model/classifier consisting M fuzzy rules, about the same operation cycle.

5.1 Scientific contribution

The scientific contribution of this work can be highlighted as follows

1. a filtering technique for time series data was developed for automatic generation of different types of feature. The suggested approach also includes an adaptation process,

2. homogeneity-oriented vector-based assignment process for overcoming the loss information and misclassification problems,

3. constriction of the size of the fuzzy rule base to the number of the related states for avoidance of the complexity and dimensionality problems,

4. statistical characteristics-based fuzzy partition process,

5. two-dimensional adaptation process for improvement of the accuracy and performance of the fuzzy rule-based systems, and

6. introduction of the modeling process framework without any type of advanced predefinition process for the modeling parameters.

5.2 Future aspects

The future work related to this work can be summarized as follows

1. reduction of the sensitivity of the proposed approach against the imbalance ration related to the data considered,

2. experimental implementation to fuzzy control systems, and

3. realization of online learning and adaptation processes.

References

[Abe01] S. Abe. *Pattern Classification: Neuro-fuzzy Methods and Their Comparison*. Springer-Verlag London Limilted, Great Britain, 2001.

[Abe10] S. Abe. *Support Vector Machines for Pattern Classification*. Springer-Verlag London Limilted, USA, 2010.

[AFAH07] J. Alcalá-Fdez, R. Alcalá, and F. Herrera. A proposal for the genetic lateral tuning of linguistic fuzzy systems and its interaction with rule selection. *IEEE Transactions on Power Delivery*, 15(4):616–635, 2007.

[AFAH11] J. Alcalá-Fdez, R. Alcalá, and F. Herrera. A fuzzy association rule-based classification model for high-dimensional problems with genetic rule selection and lateral tuning. *IEEE Transactions on Fuzzy Systems*, 19(5):857–872, 2011.

[AFFL+11] J. Alcalá-Fdez, A. Fernandez, J. Luengo, J.and Derrac, S. García, L. Sánchez, and F. Herrera. Keel data-mining software tool: Data set repository, integration of algorithms and experimental analysis framework. *Journal of Multiple-Valued Logic and Soft Computing*, 17(2-3):255–287, 2011.

[Aiz11] I. Aizenberg. *Complex-Valued Neural Networks with Multi-Valued Neurons*. Springer-Verlag Berlin Heidelberg, USA, 2011.

[AKS04] R. Abrahart, P. Kneale, and L. See. *Neural Networks for Hydrological Modelling*. A.A.Balkema Publishers, Taylor & Francis Group plc, London, UK., 2004.

[AMG08] J. M. Alonso, L. Magdalena, and S. Guillaume. HILK: A new methodology for designing highly interpretable linguistic knowledge bases using the fuzzy logic formalism. *International Journal of Intelligent Systems*, 23:761–794, 2008.

[ANHI09] R. Alcala, Y. Nojima, F. Herrera, and H. Ishibuchi. Generating single granularity-based fuzzy classification rules for multiobjective genetic fuzzy rule selection. *IEEE International Conference on Fuzzy Systems*, 2009.

[AS06] R. Andonie and L. Sasu. Fuzzy ARTMAP with input relevances. *IEEE Transactions on Neural Networks*, 17(4):929–941, 2006.

[AS10] H. Aljoumaa and D. Söffker. Condition monitoring and classification approach based on fuzzy-filtering. *Proceedings of The World Congress on Engineering and Computer Science 2010, WCECS 2010, San Francisco, USA*, pages 503–508, 2010.

[AS11a] H. Aljoumaa and D. Söffker. Adaptive fuzzy-based approach for classification of system's states. *In: Chang, F. (Ed.): Structural Health Monitoring 2011, Stanford University, Stanford, CA, Sept. 13-15, 2011*, pages 290–297, 2011.

[AS11b] H. Aljoumaa and D. Söffker. Multi-class classification approach based on fuzzy-filtering for condition monitoring. *IAENG International Journal of Computer Science*, 38(1):66–73, 2011.

[Ass07] K. Assaleh. Extraction of fetal electrocardiogram using adaptive neuro-fuzzy inference systems. *IEEE Transactions on Biomedical Engineering*, 54(1):59–68, 2007.

[AZ08]　　P. P. Angelov and X. Zhou. Evolving fuzzy-rule-based classifiers from data streams. *IEEE Transactions on Fuzzy Systems*, 16(6):1462–1475, 2008.

[BBC10]　　F. Bovolo, L. Bruzzone, and L. Carlin. Novel technique for subpixel image classification based on support vector machine. *IEEE Transactions on Image Processing*, 19(11):2983–2999, 2010.

[BBMM11]　　S. Boquera, M. J. Bleda, J. Moya, and F. Martinez. Improving offline handwritten text recognition with hybrid HMM/ANN models. *IEEE Transactions on Pattern Analysis and Machine Intelligence*, 33(4):767–779, 2011.

[Bis06]　　C. M. Bishop. *Pattern Recognition and Machine Learning*. Springer Science & Business Media, LLC, New York, NY, USA, 2006.

[BLDC09]　　d. J. M. Bezerra, A. M. N. Lima, G. S. Deep, and d. E. G. Costa. An evaluation of alternative techniques for monitoring insulator pollution. *IEEE Transactions on Power Delivery*, 24(4):1773–1780, 2009.

[Boc10]　　N. Boccara. *Modeling Complex Systems*. Springer Science & Business Media, LLC, New York, NY, 2010.

[BPB+06]　　A. Baraldi, V. Puzzolo, P. Blonda, L. Bruzzone, and C. Tarantino. Automatic spectral rule-based preliminary mapping of calibrated landsat tm and ETM+ images. *IEEE Transactions on Fuzzy Systems*, 44(9):2563–2586, 2006.

[BWGT11]　　A. Bulling, J. A. Ward, H. Gellersen, and G. Tröster. Eye movement analysis for activity recognition using electrooculography. *IEEE Transactions on Pattern Analysis and Machine Intelligence*, 33(4):741–753, 2011.

[CKLS07]　　M. Cheriet, N. Kharma, C. Liu, and C. Y. Suen. *CHARACTER RECOGNITION SYSTEMS: A Guide for Students and Practioners*. John Wiley & Sons, Canada, 2007.

[CL08]　　J. Chiang and T. M. Lee. Silico prediction of human protein interactions using fuzzy-SVM mixture models and its application to cancer research. *IEEE Transactions on Fuzzy Systems*, 16(4):1087–1095, 2008.

[CLZY11]　　W. Cai, G. Lee, M. E. Zalis, and H. Yoshida. Mosaic decomposition: An electronic cleansing method for inhomogeneously tagged regions in noncathartic CT colonography. *IEEE Transactions on Medical Imaging*, 30(3):559–574, 2011.

[CM07]　　V. CherkasskY and F. Mulier. *LEARNING FROM DATA: Concepts, Theory, and Methods*. John Wiley & Sons, Inc, Canada, 2007.

[CMP+10]　　L. Chisci, A. Mavino, G. Perferi, M. Sciandrone, C. Anile, G. Colicchio, and F. Fuggetta. Real-Time epileptic seizure prediction using ar models and support vector machines. *IEEE Transactions on Biomedical Engineering*, 57(5):1124–1132, 2010.

[Coy09]　　D. Coyle. Neural network based auto association and time-series prediction for biosignal processing in brain-computer interfaces. *IEEE Computational Intelligence Magazine*, 2009.

[CP01]　　G. Chen and T. Pham. *Fuzzy Sets,Fuzzy Logic,and Fuzzy Control Systems*. CRC Press LLC, Florida,USA, 2001.

References

[CPM09] D. Coyle, G. Prasad, and T. M. McGinnity. Faster self-organizing fuzzy neural network training and a hyperparameter analysis for a brain-computer interface. *IEEE Transactions on Systems, Man, and Cybernetics-Part B: Cybernetics*, 39(6):1458–1471, 2009.

[CPW+11] W. A. Chaovalitwongse, R. S. Pottenger, S. Wang, J. Fan, and L. D. Iasemidis. Pattern- and network-based classification techniques for multichannel medical data signals to improve brain diagnosis. *IEEE Transactions on Systems, Man, and Cybernetics-Part A: Systems and Humans*, 41(5):877–988, 2011.

[CYAP07] Y. Chen, B. Yang, A. Abraham, and L. Peng. Automatic design of hierarchical takagi-sugeno type fuzzy systems using evolutionary algorithms. *IEEE Transactions on Fuzzy Systems*, 15(3):385–397, 2007.

[DBM07] D. Dancey, Z. A. Bandar, and D. McLean. Logistic model tree extraction from artificial neural networks. *IEEE Transactions on Systems, Man, and Cybernetics-Part B: Cybernetics*, 37(4):794–802, 2007.

[Det11] K. Dettmann. *Probabilistic-based method for realizing safe and reliable mechatronic systems*. Dissertation, University of Duisburg-Essen, $http://duepublico.uni-duisburg-essen.de/servlets/DerivateServlet/Derivate-29418/DettmannKaiUweDiss.pdf$, 2011.

[DG11] N. H. Dardas and N. D. Georganas. Real-time hand gesture detection and recognition using bag-of-features and support vector machine techniques. *IEEE Transactions on Instrumentation and Measurement*, 60(11):3592–3607, 2011.

[DL07] R. Decker and H-J. Lenz. *Advances in Data Analysis*. Springer-Verlag Berlin Heidelberg, Heidelberg, 2007.

[dVB07] J. P. de Vos and M. M. Blanckenberg. Automated pediatric cardiac auscultation. *IEEE Transactions on Biomedical Engineering*, 54(2):244–252, 2007.

[DZ07] Q. Ding and N. Zhang. Classification of Recorded Musical Instruments Sounds Based on Neural Networks. *IEEE Symposium on Computational Intelligence in Image and Signal Processing (CIISP'07)*, 2007.

[EFR09] B. M. Ebrahimi, J. Faiz, and M. J. Roshtkhari. Static-, dynamic-, and mixed-eccentricity fault diagnoses in permanent-magnet synchronous motors. *IEEE Transactions on Industrial Electronics*, 56(11):4727–4739, 2009.

[Ert11] W. Ertel. *Introduction to Artificial Intelligence*. Springer-Verlag London Limited, UK, 2011.

[EVW05] J. Espinosa, J. Vandewalle, and V. Wertz. *Fuzzy Logic, Identification and Predictive Control*. Springer-Verlag London Limited, USA, 2005.

[Fen10] G. Feng. *Analysis and Synthesis of Fuzzy Control Systems A Model-Based Approach*. Taylor and Francis Group, LLC, USA, 2010.

[FKZ10] Z. Fu, A. Kelly, and J. Zhou. Mixing linear SVMs for nonlinear classification. *IEEE Transactions on Neural Networks*, 21(12):1963–1975, 2010.

[GGNZ06] I. Guyon, S. Gunn, M. Nikravesh, and L. Zadeh. *Feature Extraction, Foundations and Applications*. Springer-Verlag Berlin Heidelberg, New York, 2006.

[GJN05] H. Guo, L. B. Jack, and A. K. Nandi. Feature Generation Using Genetic Programming With Application to Fault Classification. *IEEE Transaction on Systems, Man, and Cybernetics-Part B: Cybernetics*, 35(1):89–99, 2005.

[GLQ09] H. Goh, J. Lim, and C. Quek. Fuzzy associative conjuncted maps network. *IEEE Transactions on Neural Networks*, 20(8):1302–1319, 2009.

[GÜ07] I. Güler and E. D. Übeyli. Multiclass support vector machines for EEG-signals classification. *IEEE Transactions on Information Technology in Biomedicine*, 11(2):117–126, 2007.

[HAM09] L. HAMEL. *KNOWLEDGE DISCOVERY WITH SUPPORT VECTOR MACHINES*. John-Wiley & Sons, Inc., Canada, 2009.

[HDRT04] F. Heijden, R. Duin, D. Ridder, and D. Tax. *Classification,Parameter Estimation and State Estimation: An Engineering Approach using MATLAB*. John Wiley & Sons Ltd, England, 2004.

[HH09] J. C. Hühn and E Hüllermeier. FR3: A fuzzy rule learner for inducing reliable classifiers. *IEEE Transactions on Fuzzy Systems*, 17(1):138–149, 2009.

[HHE+06] C. M. Held, J. E. Heiss, P. A. Estevez, C. A. Perez, M. Garrido, C. Algarin, and P. Peirano. Extracting fuzzy rules from polysomnographic recordings for infant sleep classification. *IEEE Transactions on Biomedical Engineering*, 53(10):1954–1962, 2006.

[HLK11] C. Huang, C. Lin, and C. Kuo. Chaos synchronization-based detector for power-quality disturbances classification in a power system. *IEEE Transactions on Power Delivery*, 26(2):944–953, 2011.

[Hut92] I. M. Hutchings. *Tribology: friction and wear of engineering materials*. CRC Press LLC, UK, 1992.

[HZH+11] H. Huang, F. Zhang, L. J. Hargrove, Z. Dou, D. R. Rogers, and K. B. Englehart. Continous locomotion-mode identification for prosthetic based on neuromuscular-mechanical fusion. *IEEE Transactions on Biomedical Engineering*, 58(10):2867–2875, 2011.

[Ibr04] A. Ibrahim. *FUZZY LOGIC for Embedded Systems Applications*. Elsevier Science, USA, 2004.

[IKN08a] H. Ishibuchi, Y. Kaisho, and Y. Nojima. Designing fuzzy rule-based classifiers that can visually explain their classification results to human users. *3rd International Workshop on Genetic and Evolving Fuzzy Systems*, 2008.

[IKN08b] H. Ishibuchi, Y. Kaisho, and Y. Nojima. A visual explanation system for explaining fuzzy reasoning results by fuzzy rule-based classifiers. *Fuzzy Information Processing Society, Annual Meeting of the North American*, 2008.

[IMT95] H. Ishibuchi, T. Murata, and I. B. Turksen. Selecting linguistic classification rules by two-objective genetic algorithms. *Proceedings of 1995 IEEE International Conference on Systems, Man and Cybernetics, Vancouver, Canada*, 1995.

[IN96] H. Ishibuchi and M. Nii. Generating fuzzy if-then rules from trained neural networks: linguistic analysis of neural networks. *IEEE International Conference on Neural Network*, 2:1133–1138, 1996.

References

[IN97] H. Ishibuchi and T. Nakashima. Performance evaluation of various variants of fuzzy classifier systems for pattern classification problems. *Fuzzy Information Processing Society, Annual Meeting of the North American*, 1997.

[IN11] H. Ishibuchi and Y. Nojima. Toward quantitative definition of explanation ability of fuzzy rule-based classifiers. *IEEE International Conference on Fuzzy Systems*, 2011.

[INK06] H. Ishibuchi, Y. Nojima, and I. Kuwajima. Fuzzy data mining by heuristic rule extraction and multiobjective genetic rule selection. *IEEE International Conference on Fuzzy Systems Sheraton Vancouver Wall Centre Hotel, Vancouver, BC, Canada*, 2006.

[INN10] H. Ishibuchi, Y. Nakashima, and Y. Nojima. Effects of fine fuzzy partitions on the generalization ability of evolutionary multi-objective fuzzy rule-based classifierss. *IEEE International Conference on Fuzzy Systems*, 2010.

[INT92] H. Ishibuchi, K. Nozaki, and H. Tanaka. Distributed representation of fuzzy rule and its application to pattern classification. *IEEE Transactions on Fuzzy Sets and Systems*, 52:21–32, 1992.

[INT95] H. Ishibuchi, T. Nakashima, and M. Tadahiko. A fuzzy classifier system that generates linguistic rules for pattern classification problems. *Proc. 2nd ICEC*, 1995.

[INT97] H. Ishibuchi, T. Nakashima, and M. Tadahiko. Simple fuzzy rule-based classification systems perform well on commonly used real-world data sets. *Fuzzy Information Processing Society, 1997 Annual Meeting of the North American*, 1997.

[INT99] H. Ishibuchi, M. Nii, and K. Tanaka. Decreasing excess fuzziness in fuzzy outputs from neural networks for linguistic rule extraction. *IEEE International Conference on Neural Network*, 6:4217–4222, 1999.

[INYT94a] H. Ishibuchi, K. Nozaki, N. Yamamoto, and H. Tanaka. Acquisition of fuzzy classification knowledge using genetic algorithms. *IEEE Transactions on Fuzzy Sets and Systems*, 3:1963–1968, 1994.

[INYT94b] H. Ishibuchi, K. Nozaki, N. Yamamoto, and H. Tanaka. Construction of fuzzy classification systems with rectangular fuzzy rules using genetic algorithms. *IEEE Transactions on Fuzzy Sets and Systems*, 65:237–253, 1994.

[INYT95] H. Ishibuchi, K. Nozaki, N. Yamamoto, and H. Tanaka. Selecting fuzzy if-then rules for classification problems using genetic algorithms. *IEEE Transactions on Fuzzy Sets and Systems*, 3(3):260–270, 1995.

[Ish07] H. Ishibuchi. Evolutionary multiobjective design of fuzzy rule-based systems. *Proceedings of the 2007 IEEE Symposium on Foundations of Computational Intelligence*, 2007.

[IY05] H. Ishibuchi and T. Yamamoto. Rule weight specification in fuzzy rule-based classification systems. *IEEE Transactions on Fuzzy Systems*, 13(4):428–435, 2005.

[Ize08] A. J. Izenman. *Modern Multivariate Statistical Techniques: Regression, Classification, and Manifold Learning*. Springer Science & Business Media, LLC, USA, 2008.

[Jam97] M. Jamshidi. *Large-Scale Systems: Modeling, Control and Fuzzy Logic*. Prentice Hall PTR, USA, 1997.

[JC07]   C. Juang and C. Chang. Human body posture classification by a neural fuzzy network and home care system application. *IEEE Transactions on Systems, Man, and Cybernetics-Part A: Systems and Humans*, 37(6):984–994, 2007.

[JCS07]   Y. Juang, S. Chiu, and S. Shiu. Fuzzy system learned through fuzzy clustering and support vector machine for human skin color segmentation. *IEEE Transactions on Systems, Man, and Cybernetics-Part A: Systems and Humans*, 37(6):1077–1087, 2007.

[JGS+11] X. Jin, Y. Guo, S. Sarkar, A. Ray, and R. M. Edwards. Anomaly detection in nuclear power plants via symbolic dynamic filtering. *IEEE Transactions on Nuclear Science*, 58(1):277–288, 2011.

[JMNR09] J. Jianmin Ma, M. N. Nguyen, and J. C. Rajapakse. Gene classification using codon usage and support vector machines. *IEEE/ACM Transactions on Computational Biology and Bioinformatics*, 6(1):134–143, 2009.

[KA06]   M. Kaya and R. Alhajj. Utilizing genetic algorithms to optimize membership functions for fuzzy weighted association rules mining. *Applied Intelligence*, 24:7–15, 2006.

[KB06]   Z. Kovačić and S. Bogdan. *Fuzzy Controller Design, Theory and Applications*. Taylor and Franis Group, LLc, 2006.

[KGS07]   T. Kulakowski, J. Gardner, and J. L. Shearer. *DYNAMIC MODELING AND CONTROL OF ENGINEERING SYSTEMS*. Cambridge University Press, New York, 2007.

[KK07]   V. G. Kaburlasos and A. Kehagias. Novel fuzzy inference system (FIS) analysis and design based on lattice theory. *IIEEE Transactions on Fuzzy Systems*, 15(2):243–260, 2007.

[KMN09]   A. Kampouraki, G. Manis, and C. Nikou. Heartbeat time series classification with support vector machines. *IEEE Transactions on Information Technology in Biomedicine*, 13(4):512–518, 2009.

[KN07]   A Klose and A. Nürnberger. On the properties of prototype-based fuzzy classifiers. *IEEE Transactions on Systems, Man, and Cybernetics-Part B: Cybernetics*, 37(4):817–835, 2007.

[Kon00]   A. Konar. *Artificial Intelligence and Soft Computing: Behavioral and Cognitive Modeling of the Human Brain*. CRC Press LLC, USA, 2000.

[KS10]   Y. U. Khan and F. Sepulveda. Brain-computer interface for single-trial EEG classification for wrist movement imagery using spatial filtering in the gamma band. *IET Signal Processing*, 4(5):510–517, 2010.

[Kun04]   L. I. Kuncheva. *Combining Pattern Classifiers, Methods and Algorithms*. John Wiley & Sons, USA, 2004.

[KXWR06] E. Keogh, X. Xi, L. Wei, and C. A. (2006) Ratanamahatana. The UCR time series classification/clustering homepage. $http://www.cs.ucr.edu/eamonn/timeseriesdata/$, 2006.

[LC07]   Y. Liu and Y. Chen. Face recognition using total margin-based adaptive fuzzy support vector machines. *IEEE Transactions on Neural Networks*, 18(1):178–192, 2007.

[LCG11]   A. Lemos, W. Caminhas, and F. Gomide. Multivariable gaussian evolving fuzzy modeling system. *IEEE Transactions on Fuzzy Systems*, 19(1):91–104, 2011.

[LFW09] C. H. Lo, E. H. K. Fung, and Y. K. Wong. Intelligent automatic fault detection for actuator failures in aircraft. *IEEE Transactions on Industrial Informatics*, 5(1):50–55, 2009.

[LHW07] Y. Liu, H. Huang, and H. Weng. Recognition of electromyographic signals using cascaded kernel learning machine. *IEEE/ASME Transactions on Mechatronics*, 12(3):253–264, 2007.

[Lin08] H. Lin. Identification of spinal deformity classification with total curvature analysis and artificial neural network. *IEEE Transactions on Biomedical Engineering*, 55(1):376–382, 2008.

[LLL+11] J. D. Londono, J. I. Llorente, N. Lechon, V. Ruiz, and G. Dominguez. Automatic detection of pathological voices using complexity measures, noise parameters, and mel-cepstral coefficients. *IEEE Transactions on Biomedical Engineering*, 58(2):370–379, 2011.

[LM08] H. Liu and H. Motoda. *Computational Methods of Feature Selection*. Taylor & Francis Group, LLC, USA, 2008.

[LWG10] J. Liu, W. Senior Wang, and F. Golnaraghi. An enhanced diagnostic scheme for bearing condition monitoring. *IEEE Transactions on Instrumentation and Measurement*, 59(2):309–321, 2010.

[LXTL11] Y. Liu, D. Xu, W. I. Tsang, and J. Luo. Textual query of personal photos facilitated by large-scale web data. *IEEE Transactions on Pattern Analysis and Machine Intelligence*, 33(5):1022–1036, 2011.

[LYL+11] K. A. Lee, C. H. You, H. Li, T. Kinnunen, and K. C. Sim. Using discrete probabilities with bhattacharyya measure for SVM-based speaker verification. *IEEE Transactions on Audio, Speech, and Language Processing*, 19(4):861–870, 2011.

[LYWL07] X. Liao, D. Yao, D. Wu, and C. Li. Combining spatial filters for the classification of single-trial EEG in a finger movement task. *IEEE Transactions on Biomedical Engineering*, 54(5):821–831, 2007.

[MASR09] F. Moreno, J. Alarcon, R. Salvador, and T. Riesgo. Reconfigurable hardware architecture of a shape recognition system based on specialized tiny neural networks with online training. *IEEE Transactions on Industrial Electronics*, 56(8):3253–3263, 2009.

[MB08] F. Melgani and Y. Bazi. Classification of electrocardiogram signals with support vector machines and particle swarm optimization. *IEEE Transactions on Information Technology in Biomedicine*, 12(5):667–677, 2008.

[MD11] M. N. Murty and V. S. Devi. *Pattern Recognition, An Algorithmic Approach*. Springer Science & Universities Press (India) Pvt. Ltd., USA, 2011.

[MMS11] W. L. Martinez, A. R. Martinez, and J. L. Solka. *Exploratory Data Analysis with MATLAB*. Taylor and Francis Group, LLC, UK, 2011.

[Moe06] F. Moens. *Information Extraction: Algorithms and Prospects in a Retrieval Context*. Springer-Verlag London Limilted, Netherlands, 2006.

[MP08a] L. Maddalena and A. Petrosino. Efficient object recognition using boundary representation and wavelet neural network. *IEEE Transactions on Neural Networks*, 19(12):2132–2149, 2008.

[MP08b] L. Maddalena and A. Petrosino. A self-organizing approach to background subtraction for visual surveillance applications. *IEEE Transactions on Image Processing*, 17(7):1168–1177, 2008.

[MR10] P. A. Mundra and J. C. Rajapakse. SVM-RFE with mrmr filter for gene selection. *IEEE Transactions on NanoBioscience*, 9(1):31–37, 2010.

[MSG07] S. K. Meher, B. U. Shankar, and A. Ghosh. Wavelet-feature-based classifiers for multispectral remote-sensing images. *IEEE Transactions on Geoscience and Remote Sensing*, 45(6):1881–1886, 2007.

[MSM09] R. A. Mohamad, L. Sulem, and C. Mokbel. Combining slanted-frame classifiers for improved HMM-based arabic handwriting recognition. *IEEE Transactions on Pattern Analysis and Machine Intelligence*, 31(7):1165–1177, 2009.

[MSR09] M. I. Murguia, J. I. Santana, and R. Rodriguez. Multiblob cosmetic defect description/classification using a fuzzy hierarchical classifier. *IEEE Transactions on Industrial Electronics*, 56(4):1292–1299, 2009.

[MTA+08] N. E. Mitrakis, C. A. Topaloglou, T. K. Alexandridis, J. B. Theocharis, and G. C. Zalidis. Decision fusion of GA self-organizing neuro-fuzzy multilayered classifiers for land cover classification using textural and spectral features. *IEEE Transactions on Geoscience and Remote Sensing*, 46(7):2137–2152, 2008.

[MZ11] F. A. Mianji and Y. Zhang. SVM-based unmixing-to-classification conversion for hyperspectral abundance quantification. *IEEE Transactions on Geoscience and Remote Sensing*, 49(11):4318–4327, 2011.

[MZK08] E. G. Mansoori, M. J. Zolghadri, and S. D. Katebi. SGERD: A steady-state genetic algorithm for extracting fuzzy classification rules from data. *IEEE Transactions on Fuzzy Systems*, 16(4):1061–1071, 2008.

[MZK09] E. G. Mansoori, M. J. Zolghadri, and S. D. Katebi. Protein superfamily classification using fuzzy rule-based classifier. *IEEE Transactions on NanoBioscience*, 8(1):92–99, 2009.

[NA08] M. S. Nixon and A. S. Aguado. *Feature extraction and Image processing*. Elsevier Ltd., UK, 2008.

[NB07] A. V. Nandedkar and P. K. Biswas. A fuzzy Min-Max neural network classifier with compensatory neuron architecture. *IEEE Transactions on Neural Networks*, 18(1):42–54, 2007.

[NB09] A. V. Nandedkar and P. K. Biswas. A granular reflex fuzzy Min-Max neural network for classification. *IEEE Transactions on Neural Networks*, 20(7):1117–1134, 2009.

[NDS+09] M. Najmabadi, V. K. Devabhaktuni, M. Sawan, S. Mayrand, and Fallone C. A. A new approach to analysis and modeling of esophageal manometry data in humans. *IEEE Transactions on Biomedical Engineering*, 56(7):1821–1830, 2009.

[NI09a] Y. Nojima and H. Ishibuchi. Effects of data reduction on the generalization ability of parallel distributed genetic fuzzy rule selection. *Ninth International Conference on Intelligent Systems Design and Applications*, 2009.

References

[NI09b] Y. Nojima and H. Ishibuchi. Interactive genetic fuzzy rule selection through evolutionary multiobjective optimization with user preference. *IEEE Symposium on Computational intelligence in miulti-criteria decision-making*, 2009.

[NIT94] K. Nozaki, H. Ishibuchi, and H. Tanaka. Trainable fuzzy classification systems based on fuzzy if-then rules. *IEEE Transactions on Fuzzy Sets and Systems*, 1:498–502, 1994.

[NIT98] T. Nakashima, H. Ishibuchi, and M. Tadahiko. Evolutionary algorithms for constructing linguistic rule-based systems for high-dimensional pattern classification problems. *IEEE World Congress on Computational Intelligence., The 1998 IEEE International Conference on Evolutionary Computation Proceedings*, 1998.

[NKI10] Y. Nojima, Y. Kaisho, and H. Ishibuchi. Accuracy improvement of genetic fuzzy rule selection with candidate rule addition and membership tuning. *Proceedings of the IEEE International Conference on Fuzzy Systems*, 2010.

[NMI10] Y. Nojima, S. Mihara, and H. Ishibuchi. Ensemble classifier design by parallel distributed implementation of genetic fuzzy rule selection for large data sets. *IEEE Congress on Evolutionary Computation*, 2010.

[NNI02] T. Nakashima, G. Nakai, and H. Ishibuchi. Improving the performance of fuzzy classification systems by membership function learning and feature selection. *Proceedings of the IEEE International Conference on Fuzzy Systems*, 1:488–493, 2002.

[NNI09] Y. Nojima, Y. Nakashima, and H. Ishibuchi. Effects of the use of multiple fuzzy partitions on the search ability of multiobjective fuzzy genetics-based machine learning. *International Conference of Soft Computing and Pattern Recognition*, 2009.

[NNI10] S. Nishikawa, Y. Nojima, and H. Ishibuchi. Appropriate granularity specification for fuzzy classifier design by data complexity measures. *Second World Congress on Nature and Biologically Inspired Computing*, 2010.

[NNI11] Y. Nojima, S. Nishikawa, and H. Ishibuchi. A meta-fuzzy classifier for specifying appropriate fuzzy partitions by genetic fuzzy rule selection with data complexity measures. *IEEE International Conference on Fuzzy Systems*, 2011.

[Now08] R. Nowicki. On combining neuro-fuzzy architectures with the rough set theory to solve classification problems with incomplete data. *IEEE Transactions on Knowledge and Data Engineering*, 20(9):1239–1253, 2008.

[NTI97] T. Nakashima, M. Tadahiko, and H. Ishibuchi. Input selection in fuzzy rule-based classification systems. *Proceedings of the Sixth IEEE International Conference on Fuzzy Systems*, 3:1457–1462, 1997.

[NYSI06] T. Nakashima, Y. Yokota, G. Schaefer, and H. Ishibuchi. A cost-based fuzzy rule-based system for pattern classification problems. *IEEE International Conference on Fuzzy Systems Sheraton Vancouver Wall Centre Hotel, Vancouver, BC, Canada*, 2006.

[OBC06] O. Ondel, E. Boutleux, and G. Clerc. A Method to Detect Broken Bars in Induction Machine Using Pattern Recognition Techniques. *IEEE Transaction on Industry Applications*, 42(4):916–923, 2006.

[OD08]   D. L. Olson and D. Delen. *Advanced Data Mining Techniques*. Springer-Verlag Berlin Heidelberg, German, 2008.

[OH08]   M. A. Oskoei and H. Hu. Support vector machine-based classification scheme for myoelectric control applied to upper limb. *IEEE Transactions on Biomedical Engineering*, 55(8):1956–1965, 2008.

[OTS10]   Y. O. Ouma, R. Tateishi, and J. T. Sumantyo. Urban features recognition and extraction from very-high resolution multi-spectral satellite imagery: a micro-macro texture determination and integration framework. *IET Image Processing*, 4(4):235–254, 2010.

[PCM09]   A. Puig, J. Casillas, and E. Mansilla. Fuzzy-ucs: A michigan-style learning fuzzy-classifier system for supervised learning. *IEEE Transactions on Evolutionary Computation*, 13(2):260–383, 2009.

[PH10a]   C. Petrantonakis and L. J. Hadjileontiadis. Emotion recognition from brain signals using hybrid adaptive filtering and higher order crossings analysis. *IEEE Transactions on Affective Computing*, 1(2):81–97, 2010.

[PH10b]   C. Petrantonakis and L. J. Hadjileontiadis. Emotion recognition from eeg using higher order crossings. *IEEE Transactions on Information Technology in Biomedicine*, 14(2):186–197, 2010.

[PH11]   P. C. Petrantonakis and L. J. Hadjileontiadis. A novel emotion elicitation index using frontal brain asymmetry for enhanced EEG-based emotion recognition. *IEEE Transactions on Information Technology in Biomedicine*, 15(5):737–746, 2011.

[PJL$^+$06] R. J. Povinelli, M. T. Johnson, A. C. Lindgren, F. M. Roberts, and J. Ye. Statistical models of reconstructed phase spaces for signal classification. *IEEE Transactions on Signal Processing*, 54(6):2178–2186, 2006.

[PP09]   B. K. Panigrahi and V. R. Pandi. Optimal feature selection for classification of power quality disturbances using wavelet packet-based fuzzy k-nearest neighbour algorithm. *IET Generation, Transmission & Distribution*, 3(3):296–306, 2009.

[QLT10]   A. Quteishat, C. P. Lim, and K. S. Tan. A modified Fuzzy Min-Max neural network with a genetic-algorithm-based rule extractor for pattern classification. *IEEE Transactions on Systems, Man, and Cybernetics-Part A: Systems and Humans*, 40(3):641–650, 2010.

[RACS11]   J. M. T. Romano, R. R. F. Attux, C. C. Cavalcante, and R. Suyama. *UNSUPERVISED SIGNAL PROCESSING, Channel Equalization and Source Separation*. Taylor and Francis Group, LLC, UK, 2011.

[RAT11]   M. M. Rahman, S. K. Antani, and G. R. Thoma. A learning-based similarity fusion and filtering approach for biomedical image retrieval using SVM classification and relevance feedback. *IEEE Transactions on Information Technology in Biomedicine*, 15(4):640–646, 2011.

[Ros10]   T. J. Ross. *Fuzzy Logic WITH Engineering Applications*. John Wiley & Sons Ltd, Chichester, England, 2010.

[Rut04]   L. Rutkowski. *FLEXIBLE NEURO-FUZZY SYSTEMS: Structures, Learning and Performance Evaluation*. Kluwer Academic Publishers, USA, 2004.

References

[SAC07] Z. Sun, K. Au, and T. Choi. A neuro-fuzzy inference system through integration of fuzzy logic and extreme learning machines. *IEEE Transactions on Systems, Man, and Cybernetics-Part B: Cybernetics*, 37(5):1321–1331, 2007.

[SAK11] H. Sahbi, Y. Audibert, and R. Keriven. Context-dependent kernels for object classification. *IEEE Transactions on Pattern Analysis and Machine Intelligence*, 33(4):699–708, 2011.

[Sam07] S. Samarasinghe. *Neural Networks for Applied Sciences and Engineering From Fundamentals to Complex Pattern Recognition*. Taylor & Francis Group, LLC, UK, 2007.

[SAS08] D. Söffker, H. Aljoumaa, and M. Saadawia. Defining Features for Diagnosis and Prognosis - Part II: Data Driven Adaption of Diagnosis Filter. *International Conference on Motion and Vibration Control (MOVIC)*, 2008.

[SEAJK10] S. R. Samantaray, K. El-Arroudi, G. Joos, and I. Kamwa. A fuzzy rule-based approach for islanding detection in distributed generation. *IEEE Transactions on Power Delivery*, 25(3):1427–1433, 2010.

[SGT+09] F. Scarselli, M. Gori, A. C. Tsoi, M. Hagenbuchner, and G. Monfardini. The graph neural network model. *IEEE Transactions on Neural Networks*, 20(1):61–80, 2009.

[SIL07] Y. Saeys, I. Inza, and P. Larranaga. A review of feature selection techniques in bioinformatics. *Bioinformatics Advance Access*, 23(19):2507–2517, 2007.

[SMW+11] H. J. Song, C. Y. Miao, R. Wuyts, , Z. Q. Shen, M. D'Hondt, and F. Catthoor. An extension to fuzzy cognitive maps for classification and prediction. *IEEE Transactions on Fuzzy Systems*, 19(1):116–135, 2011.

[SN10] G. Schaefer and T. Nakashima. Data mining of gene expression data by fuzzy and hybrid fuzzy methods. *IEEE Transactions on Information Technology in Biomedicine*, 14(1):23–29, 2010.

[SNYI07] G. Schaefer, T. Nakashima, Y. Yokota, and H. Ishibuchi. Cost-sensitive fuzzy classification for medical diagnosis. *Proceedings of the 2007 IEEE Symposium on Computational Intelligence in Bioinformatics and Computational Biology*, 2007.

[SS10] S. Sumathi and P. Surekha. *Computational Intelligence Paradigms Theory and Applications using MATLAB®*. Taylor and Franis Group, LLc, 2010.

[SSB06] K. M. Silva, B. A. Souza, and N. S. D. Brito. Fault detection and classification in transmission lines based on wavelet transform and ANN. *IEEE Transactions on Power Delivery*, 21(4):2058–2063, 2006.

[SSD07] N. Sivanandam, S., S. Sumathi, and S. N. Deepa. *Introduction to Fuzzy Logic using MATLAB*. Springer-Verlag Berlin Heidelberg, German, 2007.

[STS11] M. Spigai, C. Tison, and J. Souyris. Time-Frequency analysis in high-resolution SAR imagery. *IEEE Transactions on Geoscience and Remote Sensing*, 49(7):2699–2711, 2011.

[SWSC07] R. Shuyler, A. White, K. Staley, and K. J. Cios. Identification of ictal and pre-ictal states using rbf networks with wavelet-decomposed EEG data. *IEEE Engineering in Medicine and Biology Magazine*, 2007.

[TGA+11] N. N. Tsiaparas, S. Golemati, I. Andreadis, J. S. Stoitsis, I. Valavanis, and Nikita S. N. Comparison of multiresolution features for texture classification of carotid atherosclerosis from b-mode ultrasound. *IEEE Transactions on Information Technology in Biomedicine*, 15(1):130–137, 2011.

[TK09] S. Theodoridis and K. Koutroumbas. *Pattern Recognition*. Elsevier Inc., USA, 2009.

[TL11] H. D. Tran and H. Li. Sound event recognition with probabilistic distance SVMs. *IEEE Transactions on Audio, Speech, and Language Processing*, 19(6):1556–1568, 2011.

[TNQ08] T. Z. Tan, G. S. Ng, and C. Quek. A novel biologically and psychologically inspired fuzzy decision support system: Hierarchical complementary learning. *IEEE/ACM Transactions on Computational Biology and Bioinformatics*, 5(1):67–79, 2008.

[TZCK09] Y. Tang, Q. Zhang, N. V. Chawla, and S. Krasser. SVMs modeling for highly imbalanced classification. *IEEE Transactions on Systems, Man, and Cybernetics-Part B: Cybernetics*, 39(1):281–288, 2009.

[UGSR10] J. Upendar, C. P. Gupta1, G. K. Singh1, and G. Ramakrishna. PSO and ANN-based fault classification for protective relaying. *IET Generation, Transmission & Distribution*, 4(10):1197–1212, 2010.

[Vap95] N. Vapnik. *The Nature of Statistical Learning Theory*. Springer, New York, 1995.

[Vap98] N. Vapnik. *Statistical Learning Theory*. John Wiley & Sons Ltd, Canada, 1998.

[VMM10] F. Veredas, H. Mesa, and L. Morente. Binary tissue classification on wound images with neural networks and bayesian classifiers. *IEEE Transactions on Medical Imaging*, 29(2):410–427, 2010.

[VZ09] M. Varma and A. Zisserman. A statistical approach to material classification using image patch exemplars. *IEEE Transactions on Pattern Analysis and Machine Intelligence*, 31(11):2032–2047, 2009.

[Wan08a] W. Wang. An enhanced diagnostic system for gear system monitoring. *IEEE Transactions on Systems, Man, and Cybernetics-Part B: Cybernetics*, 38(1):102–112, 2008.

[Wan08b] W. Wang. An intelligent system for machinery condition monitoring. *IEEE Transactions on Fuzzy Systems*, 16(1):110–122, 2008.

[WC11] C. Wang and T. Chen. Rapid detection of small oscillation faults via deterministic learning. *IEEE Transactions on Neural Networks*, 22(8):1284–1296, 2011.

[WF05] H. Witten and E. Frank. *Data Mining: Practical Machine Learning Tools and Techniques*. Morgan Kaufmann Publishers, USA, 2005.

[WH06] J. Wang and J. Hao. A new version of 2-Tuple fuzzy linguistic representation model for computing with words. *IEEE Transactions on Fuzzy Systems*, 14(3):435–445, 2006.

[WI11] B. M. Wilamowski and J. I. Irwin. *IntellIgent systems*. Taylor and Francis Group, LLC, USA, 2011.

[WK09] X. Wu and V. Kummar. *The Top Ten Algorithms in Data Mining*. Taylor & Francis Group, LLC, USA, 2009.

References

[WK10]   Y. Wu and S. Krishnan. Statistical analysis of gait rhythm in patients with parkinson's disease. *IEEE Transactions on Neural Systems and Rehabilitation Engineering*, 18(2):150–158, 2010.

[WM08]   M. Wenxing and L. Meng. Fault Pattern Recognition of Roling Bearings Based on Wavelet Packet and Support Vector Machine. *27th Chinese Control Conference*, pages 65–68, 2008.

[WMLL09] Z. Wang, A. Maier, N. K. Logothetis, and H. Liang. Extraction of bistable-percept-related features from local field potential by integration of local regression and common spatial patterns. *IEEE Transactions on Biomedical Engineering*, 56(8):2095–2103, 2009.

[WMSS10] J. J. Wade, L. J. McDaid, J. A. Santos, and H. M. Sayers. SWAT: A spiking neural network training algorithm for classification problems. *IEEE Transactions on Neural Networks*, 21(11):1817–1830, 2010.

[Wol01]   O. Wolkenhauer. *Data Engineering: Fuzzy Mathematics in Systems Theory and Data Analysis*. John Wiley & Sons, Inc, USA, 2001.

[WSYT07] D. Wang, D. S. S. Yeung, and E. C. C. Tsang. Weighted mahalanobis distance kernels for support vector machines. *IEEE Transactions on Neural Networks*, 18(3):1453–1462, 2007.

[YCSN10] D. S. Yeung, I. Cloete, D. Shi, and W. W. Y. Ng. *Sensitivity Analysis for Neural Networks*. Springer-Verlag Berlin Heidelberg, German, 2010.

[YKS07]   P. H. Yeomans, B. V. K. V. Kumar, and M. Savvides. Palmprint classification using multiple advanced correlation filters and palm-specific segmentation. *IEEE Transactions on Information Forensics and Security*, 2(3):613–622, 2007.

[YKTZ11] Y. Yin, I. Kaku, J. Tang, and J. Zhu. *Data Mining: Concepts, Methods and Applications in Management and Engineering Design*. Springer-Verlag London Limited, UK, 2011.

[YLM+07] H. Ying, F. Lin, R. D. MacArthur, J. A. Cohn, D. C. Jones, H. Ye, and L. R. Crane. A self-learning fuzzy discrete event system for HIV/AIDS treatment regimen selection. *IEEE Transactions on Systems, Man, and Cybernetics-Part B: Cybernetics*, 37(4):966–979, 2007.

[Zad65]   L.A. Zadeh. Fuzzy sets. *Information and Control*, 8(3):338–353, 1965.

[ZG07]   S. Zhou and J. Q. Gan. Constructing L2-SVM-based fuzzy classifiers in high-dimensional space with automatic model selection and fuzzy rule ranking. *IEEE Transactions on Fuzzy Systems*, 15(3):398–409, 2007.

[Zha08]   P. Zhang. *Industrial Control Technology, A Handbook for Engineers and Researchers*. William Andrew Inc., 2008.

[ZLH10]   Z. Zhang, K. D. K. Luk, and Y. Hu. Identification of detailed time-frequency components in somatosensory evoked potentials. *IEEE Transactions on Neural Systems and Rehabilitation Engineering*, 18(3):245–254, 2010.

[ZZGL07] Y. Zhong, L. Zhang, J. Gong, and P. Li. A supervised artificial immune classifier for remote-sensing imagery. *IEEE Transactions on Geoscience and Remote Sensing*, 45(12):3957–3966, 2007.

During the design and development of the Adaptive Fuzzy-based Approach, the different development steps were published in the following publications:

- Söffker, D.; Aljoumaa, H.; Baccar, D.; Rothe, S.: Smart, Robust und Einfach: Drei Innovative Konzepte zur Maschinen-Diagnose, 9. Aachener Kolloquium für Instandhaltung, Diagnose und Anlagenüberwachung AKIDA, Aachen, 2012 (accepted).

- Aljoumaa, H.; Söffker, D.: Adaptive Fuzzy-based Approach for Classification of System's States. In: Chang, F. (Ed.): Structural Health Monitoring 2011, Stanford University, Stanford, CA, Sept. 13-15, 2011, pp. 290-297, 2011.

- Aljoumaa, H.; Söffker, D.: Multi-Class Approach based on Fuzzy-Filtering for Condition Monitoring. IAENG International Journal of Computer Science, Vol. 38, Issue 1, pp. 66-73, 2011.

- Aljoumaa, H.; Söffker, D.: Condition Monitoring and Classification Approach based on Fuzzy-Filtering, Proceedings of The World Congress on Engineering and Computer Science 2010, WCECS 2010, San Francisco, USA, pp. 503-508, 2010.

- Söffker, D.; Aljoumaa, H.: Signal-based modeling - a new method for classifying system states. In: Chang, F. (Ed.): Structural Health Monitoring 2011, Stanford University, Stanford, CA, Sept. 9-11, 2009, pp. 1519-1527, 2009.

- Söffker, D.; Aljoumaa, H.; Saadawia, M.; Dettmann, K.-U.: Beschreibung des Maschinenzustandes durch komplexe Merkmale zur Schadenfrüherkennung und zum Kondition Monitoring, 7. Aachener Kolloquium für Instandhaltung, Diagnose und Anlagenüberwachung AKIDA, Aachen, S. 33-50, 2008.

- Söffker, D.; Aljoumaa, H.; Saadawia, M.: Defining features for diagnosis and prognosis - Part II: Data driven adaption of diagnosis filter. Proc. 9th International Conference on Motion and Vibration Control MOVIC 2008, Munich, 2008.

i want morebooks!

Buy your books fast and straightforward online - at one of world's fastest growing online book stores! Environmentally sound due to Print-on-Demand technologies.

Buy your books online at
www.get-morebooks.com

Kaufen Sie Ihre Bücher schnell und unkompliziert online – auf einer der am schnellsten wachsenden Buchhandelsplattformen weltweit! Dank Print-On-Demand umwelt- und ressourcenschonend produziert.

Bücher schneller online kaufen
www.morebooks.de

VDM Verlagsservicegesellschaft mbH
Heinrich-Böcking-Str. 6-8   Telefon: +49 681 3720 174   info@vdm-vsg.ce
D - 66121 Saarbrücken   Telefax: +49 681 3720 1749   www.vdm-vsg.ce

Printed by Books on Demand GmbH, Norderstedt / Germany